INQUIRY-BASED LEARNING FOR SCIENCE, TECHNOLOGY, ENGINEERING, AND MATH (STEM) PROGRAMS: A CONCEPTUAL AND PRACTICAL RESOURCE FOR EDUCATORS

INNOVATIONS IN HIGHER EDUCATION TEACHING AND LEARNING

Series Editor: Patrick Blessinger

Recent Volumes:

INNOVATIONS IN HIGHER EDUCATION TEACHING AND LEARNING VOLUME 4

INQUIRY-BASED LEARNING FOR SCIENCE, TECHNOLOGY, ENGINEERING, AND MATH (STEM) PROGRAMS: A CONCEPTUAL AND PRACTICAL RESOURCE FOR EDUCATORS

EDITED BY

PATRICK BLESSINGER
International HETL Association, New York, NY, USA

JOHN M. CARFORA
Loyola Marymount University, Los Angeles, CA, USA

Created in partnership with the
International Higher Education Teaching and Learning Association

Higher Education
Teaching & Learning

https://www.hetl.org/

Emerald

United Kingdom – North America – Japan
India – Malaysia – China

Emerald Group Publishing Limited
Howard House, Wagon Lane, Bingley BD16 1WA, UK

First edition 2015

British Library Cataloguing in Publication Data
A catalogue record for this book is available from the British Library

ISBN: 978-1-78441-850-2
ISSN: 2055-3641 (Series)

CONTENTS

PART II
PRACTICES AND STRATEGIES

LIST OF CONTRIBUTORS

Amelia J. Ahern-Rindell	University of Portland, Portland, OR, USA
Teresa Bettencourt	University of Aveiro, Aveiro, Portugal
Patrick Blessinger	International HETL Association, New York, NY, USA
Barb Bloemhof	McMaster University, Hamilton, Ontario, Canada
Jacqueline Broerse	VU University, Amsterdam, The Netherlands
Mindy Capaldi	Valparaiso University, Valparaiso, IN, USA
John M. Carfora	Loyola Marymount University, Los Angeles, CA, USA
Lorraine Carter	Nipissing University, North Bay, Ontario, Canada
Fabio Dovigo	University of Bergamo, Bergamo, Italy
Richard R. Hake	Indiana University, Bloomington, IN, USA
Philip Hawes	Whitireia Community Polytechnic, Porirua, New Zealand
Henk Huijser	Batchelor Institute of Indigenous Tertiary Education, Batchelor, Australia
Megan Y. C. A. Kek	University of Southern Queensland, Toowoomba, Australia
Les Kirkup	University of Technology, Sydney, New South Wales, Australia
Wanda S. Konijn	VU University, Amsterdam, The Netherlands

Natalie Kuldell	Massachusetts Institute of Technology, Cambridge, MA, USA
Jeroen Maas	VU University, Amsterdam, The Netherlands
Gina J. Mariano	Troy University, Troy, AL, USA
William F. McComas	University of Arkansas, Fayetteville, AR, USA
Rudolph Mitchell	Massachusetts Institute of Technology, Cambridge, MA, USA
Vanessa R. Pitts Bannister	Florida A&M University, Tallahassee, FL, USA
Vincenza Rocco	University of Bergamo, Bergamo, Italy
Richard E. Shope	Loyola Marymount University, Los Angeles, CA, USA
Susana Silva	University of Coimbra, Coimbra, Portugal
Lorraine Stefani	University of Auckland, Auckland, New Zealand
Ana Cristina Tavares	University of Coimbra, Coimbra, Portugal
Ruth Terwijn	University of Southern Queensland, Toowoomba, Australia
Wendy Trimmer	Whitireia Community Polytechnic, Porirua, New Zealand
Dylan Powell Williams	University of Leicester, Leicester, UK
Baiba Zarins	Nipissing University, North Bay, Ontario, Canada
Marjolein B. M. Zweekhorst	VU University, Amsterdam, The Netherlands

SERIES EDITOR'S INTRODUCTION

The purpose of this series is to publish current research and scholarship on innovative teaching and learning practices in higher education. The series is developed around the premise that teaching and learning is more effective when instructors and students are actively and meaningfully engaged in the teaching-learning process.

The main objectives of this series are to:

1. present how innovative teaching and learning practices are being used in higher education institutions around the world across a wide variety of disciplines and countries,
2. present the latest models, theories, concepts, paradigms, and frameworks that educators should consider when adopting, implementing, assessing, and evaluating innovative teaching and learning practices, and
3. consider the implications of theory and practice on policy, strategy, and leadership.

This series will appeal to anyone in higher education who is involved in the teaching and learning process from any discipline, institutional type, or nationality. The volumes in this series will focus on a variety of authentic case studies and other empirical research that illustrates how educators from around the world are using innovative approaches to create more effective and meaningful learning environments.

Innovation teaching and learning is any approach, strategy, method, practice, or means that has been shown to improve, enhance, or transform the teaching-learning environment. Innovation involves doing things differently or in a novel way in order to improve outcomes. In short, innovation is positive change. With respect to teaching and learning, innovation is the implementation of new or improved educational practices that result in improved educational and learning outcomes. This innovation can be any positive change related to teaching, curriculum, assessment, technology, or other tools, programs, policies, or processes that leads to improved educational and learning outcomes. Innovation can occur in institutional development, program development, professional development, or learning development.

The volumes in this series will not only highlight the benefits and theoretical frameworks of such innovations through authentic case studies and other empirical research but also look at the challenges and contexts associated with implementing and assessing innovative teaching and learning practices. The volumes represent all disciplines from a wide range of national, cultural, and organizational contexts. The volumes in this series will explore a wide variety of teaching and learning topics such as active learning, integrative learning, transformative learning, inquiry-based learning, problem-based learning, meaningful learning, blended learning, creative learning, experiential learning, lifelong and lifewide learning, global learning, learning assessment and analytics, student research, faculty and student learning communities, as well as other topics.

This series brings together distinguished scholars and educational practitioners from around the world to disseminate the latest knowledge on innovative teaching and learning scholarship and practices. The authors offer a range of disciplinary perspectives from different cultural contexts. This series provides a unique and valuable resource for instructors, administrators, and anyone interested in improving and transforming teaching and learning.

Patrick Blessinger
Founder and Executive Director,
International HETL Association

FOREWORD

In the United States in particular, millions of dollars have been poured into improving undergraduate education in science, technology, engineering and mathematics (STEM) education. This is in response to mounting concern about a lack of scientists, engineers and mathematicians to keep the United States in the forefront of research, innovation and technology. As the chapters in this book demonstrate this focus on STEM education has resulted in research into, and implementation of, a range of different classroom practices, the development of curricula designed for authentic STEM learning experiences and shifts in pedagogical models and frameworks. A prominent development has been that of inquiry-based learning (IBL) as applied to STEM education. Interestingly in Australia and the United Kingdom the terms IBL and problem-based learning (PBL) are used almost interchangeably albeit there are potential philosophical and conceptual differences between definitions of the terms.

There is a school of thought suggesting that PBL is one of the most innovative, instructional methods conceived in the history of education. As a pedagogical approach, PBL first came to prominence in response to the widespread criticism that traditional teaching and learning methods fail to prepare medical students for solving problems in clinical settings. This is not a dissimilar reason for IBL coming to the forefront in STEM education. The significant commonalities between PBL and IBL are that they are pedagogies which have their origins in the work of the American philosopher and educator John Dewey (1859–1952), father of the experiential learning movement. Dewey's view of learning essentially assigns a major role to what can be best described as a creative intelligence in the discovery of new ideas in the disciplines. At its best, IBL is a student-centred active learning approach, focusing on questioning, critical thinking and problem solving.

What makes IBL as applied to STEM education intriguing is the idea of STEM as a meta-discipline, that is, the creation of a discipline based on the integration of other disciplinary knowledge into a new 'whole'. STEM education is an inter-disciplinary, some might say trans-disciplinary approach where rigorous academic concepts are coupled with authentic real-world learning as students apply science, technology, engineering and mathematics learning that can help to make connections between formal learning, community learning and work-based situations. It is in no way an easy task to achieve a meta-disciplinary approach in learning and teaching and it would be wrong to suggest that we have completed this journey with respect to integrative STEM education. However, by applying IBL within the STEM subject areas, we are by definition: facilitating student learning rather than pushing students through a content-driven curriculum (guide on the side rather than sage on the stage); supporting the development of self-directed learning skills (inquiring, questioning, researching – all of which are likely to encourage boundary crossing between disciplines); supporting co-creation of new knowledge (empowering students to create knowledge rather than having all knowledge imposed).

Clearly, the effective use of new technologies in the classroom greatly enhances the potential of IBL in STEM education. Learning is no longer confined to official classroom hours or the availability of textbooks. Most students in the 21st century carry with them their own personal digital environments; they communicate constantly using social media and they are consummate consumers of information instantly available to them. The real trick for meaningful learning lies in strategies that enhance their information literacy skills. Although IBL predates the technologies we have available to us today, new technologies might be said to be 'made to measure' for IBL and IBL as applied to STEM in particular. Digital technologies support a range of learning styles and strategies and their huge advantage lies in the potential for connectivity and creativity as well as access to increasingly sophisticated software and experimentation in learning.

The chapters in this book will inspire and motivate all STEM teachers and facilitators of learning to take the ideas into their own classrooms, to provide authentic learning experiences for learners and to increase engagement for STEM students.

Lorraine Stefani
Professor of Higher Education Strategic Engagement
Faculty of Education, University of Auckland

PART I
CONCEPTS AND PRINCIPLES

INNOVATIVE APPROACHES IN TEACHING AND LEARNING: AN INTRODUCTION TO INQUIRY-BASED LEARNING FOR STEM PROGRAMS

Patrick Blessinger and John M. Carfora

ABSTRACT

This chapter provides an introduction to how the inquiry-based learning (IBL) approach is being used by colleges and universities around the world to strengthen the interconnections between teaching, learning, and research within STEM programs. This chapter provides a synthesis and analysis of the chapters in the volume, which present a range of case studies and empirical research on how IBL is being used across a range of courses across a range of institutions within STEM programs. Based on these findings, this chapter argues that the IBL approach has great potential to enhance and transform teaching and learning. Given the growing demands placed on education to meet a diverse range of complex political, economic, and social problems and personal needs, this chapter argues that education should be a place where students learn

Inquiry-Based Learning for Science, Technology, Engineering, and Math (STEM) Programs: A Conceptual and Practical Resource for Educators
Innovations in Higher Education Teaching and Learning, Volume 4, 3–19

ISSN: 2055-3641/doi:10.1108/S2055-364120150000004001

"how-to-learn" – where increasingly higher levels of self-directed learning is fostered – and where students grow in the three key areas of learning: affectively, behaviorally, and cognitively. To that end, this chapter argues that IBL, if designed and implemented properly, can be an important approach to enhancing and transforming teaching and learning in higher education.

Keywords: STEM; inquiry; active learning; meaningful learning; problem-based learning

INTRODUCTION

As the academic knowledge base and research has grown over the last few decades on the effectiveness of inquiry-based learning (IBL), an increasing number of colleges and universities have adopted this approach as a means to enhance and transform the teaching and learning experience for instructors and students. This volume presents research-based evidenced by educational scholars and practitioners from around the world on how IBL is being used to strengthen the relations between teaching, learning, and research. The potential for IBL enhance the teaching, learning, and research nexus lies in nurturing a mindset of inquiry that infuses all aspects of the teaching–learning experience (Carfora, 2011; Lee, 2013). Since the main purpose of any human-based organism (e.g., individual, organizational, community) is to organize and make meaning from their experiences, they naturally ask questions in order to investigate and understand their experiences (Daloz Parks, 2000; Doherty, Riordan, & Roth, 2003; Kovbasyuk & Blessinger, 2013). Thus, asking questions and seeking answers to those questions is a natural part of learning in the human experience. IBL with its focus on inquiry, intentionality, and meaning-making as a means to learn is therefore well suited for this purpose.

Educational institutions can therefore serve as catalysts to cultivate this inquiry-oriented learning process and also provide a safe and nurturing environment to question, to investigate, to take reasonable risks, and to develop more self-regulated learners. IBL can be used to foster higher-order thinking, develop deeper intellectual capacities, as well as clarify personal values, identity, and meaning-making. Mezirow (1991) states that, "Making meaning is central to what learning is all about" (p. 11). According to Mezirow (1991), people are better able to accept knowledge when it is consistent with their existing knowledge base and their own experiences. Kovbasyuk and Blessinger (2013) explain why open meaning-making educational processes are important to advanced learning because this type of learning challenges learners to investigate a question or problem from multiple perspectives. In inquiry-based learning spaces, educators help create learning spaces in which continual questioning and meaning-making is at the heart of the teaching–learning process.

In response to some of the perceived shortcomings of IBL (Kirschner, Sweller, & Clark, 2006), Hmelo-Silver, Duncan, and Chinn (2007) present a large body of evidence that shows the effectiveness of IBL as a teaching–learning strategy if designed and implemented properly relative to the particular teaching–learning context. The authors conclude that "… there is growing evidence from large-scale experimental and quasi-experimental studies demonstrating that inquiry-based instruction results in significant learning gains in comparison to traditional instruction and that disadvantaged students benefit most from inquiry-based instructional approaches" (p. 104). The empirical evidence in this volume, as well as the three previous IBL volumes in this series, adds to this IBL knowledge base. At the heart of IBL is the recognition that education is about learning both the epistemic knowledge and the epistemic practices of the domain(s) being studied since both components together help to more effectively prepare students to begin thinking like a practitioner in the domain (e.g., to think like an chemist, to think like an engineer, to think like a mathematician), which in turn produces learning that is more authentic (Bereiter & Scardamalia, 2006; Bransford, Brown, & Cocking, 2000; Sandoval & Reiser, 2004).

In light of this large knowledge base of evidence on the effectiveness of IBL, the chapters in this volume demonstrate how the IBL approach is being utilized across a range of educational sceneries and institutional types in the STEM fields. The examples provided in this volume (e.g., case studies, empirical research studies) provide in-depth and broad examples of how IBL can be used to design flexible and effective learning environments for the STEM fields. The chapters in this volume cover the core theories, concepts, and principles related to IBL as well successful practices and strategies for IBL implementations at a variety of colleges and universities around the world.

The chapters in this volume illustrate the effectiveness of IBL to work within any institutional regardless of institutional type because IBL overlaps and integrates with several important learning theories that have been shown to be effective in cultivating higher-order thinking and academic motivation among students (e.g., active learning, problem-based learning, project-based learning, meaningful learning). The chapters in this volume provide a wide- and far-reaching overview of recent developments in IBL theories and pedagogical practices (Barrow, 2006; Bell, Urhahne, Schanze, & Ploetzner, 2010; Blessinger & Carfora, 2014a, 2014b, 2015; Bruner, 1961, 1990; Eslinger, White, & Frederiksen, 2008; Garrison, Anderson, & Archer, 2000; Gredler, 2009; Hmelo-Silver et al., 2007; Lee, Greene, Odom, Schechter, & Slatta, 2004; National Research Council, 1996; Spronken-Smith, 2012; Vygotsky, 1962, 1978).

Recent developments (e.g., changing student demographics, shifting workforce needs, globalization) are requiring institutions to redefine their notions of teaching and learning in the 21st century. In response to these growing pressures, more institutions are seeking more effective ways to engage and prepare students. To this end, Kuh (2008) identified several high impact learning models that increase student engagement (e.g., service learning, civic engagement, undergraduate research, education abroad). IBL creates new opportunities for instructors, students, and institutions to teach and learn in more effective ways. This is especially important since higher education in the 21st century is a more complex enterprise than in previous decades due to increased demands placed upon it by society, policy makers, employers, and other group(s) that have an interest in the outcomes shaped by higher education. The examples provided in this volume therefore serve to illustrate on how IBL is being used in novel ways in the STEM fields.

The chapters in this volume also illustrate the dramatic impact advances in new learning research and theories are having upon educators' notions of what it means to teach and learn. The modern era is characterized by dramatic increases in the production of new knowledge – and conversely, the dramatic decrease in the shelf-life of new knowledge – and the growing democratization of higher education (Blessinger & Anchan, 2015). Thus, IBL with its inherent qualities of authenticity, real-world problem-solving, and focus on open mean-making offers a promising approach to enhance and transform the quality of teaching and learning by adopting a more learning-centered teaching approach.

ADOPTION PRINCIPLES

Two key principles emerge from the findings of the chapters in this volume that help frame on how IBL is being used within the STEM fields:

1. IBL is an approach to active learning that is driven by authentic problem-based questions. IBL is typically grounded in a constructivist based approach to teaching and learning and has its roots in the educational philosophy of John Dewey, which states that the essence of learning is the process of constructing meaning from knowledge consumption and production. Key ingredients in this process are learning by doing and developing the skill to "learn how to learn." Because IBL is often oriented around learning by doing, it often involves problem-based learning and project-based learning also which is naturally suited for STEM disciplines.
2. IBL is considered a learning-centered approach to education where the instructors serve as facilitators and guides in the inquiry process. Most inquiry-based projects are not completely unstructured. Rather, they are operated on a continuum and require appropriate planning, design, and assessment by the instructor. Thus, the roles of instructors and students are different than in traditional teaching-centered or curriculum-centered classrooms. With IBL, students learn to take increasing amounts of responsibility for their own learning (self-regulated learning) and they learn in more authentic ways (meaningful learning).

These principles reflect that IBL must be adapted to the context of the learning goals of the course and the mission, vision, and shared values of the institution. Although learners assume increasing responsibility for their own learning, the instructor plays a critical role in creating the necessary conditions for the learning environment. Since learners are learning to be more self-directed, academic motivation is likely to be increased as well, all else being equal. Moving from deeply embedded traditional teaching–learning roles (e.g., the instructor as the sole knowledge expert, the lecture as the only means to facilitate learning, the textbook as the only or main repository of information, and the student as passive recipient of information dispensed solely through a steady diet of lectures) to IBL roles can be a major transition because along with taking on new roles comes the development of a new mindset about the purpose and nature of teaching and learning in the modern era, which may be a formidable hurdle for those who are deeply entrenched in more traditional modes of teaching and learning.

While the student makes the transition from a passive participant to an active participant in the learning process, the instructor must also make the transition from solely a content specialist to an instructional leader, a learning architect, and a learning mentor and facilitator. In the IBL teaching–learning approach, both roles are greatly expanded, better defined, and better focused on attaining more explicit, higher-order learning outcomes. Also important is the quality of the relationship between instructor and learner, as well as between learners. IBL requires more interaction and collaboration among all participants. IBL requires careful planning and a great deal of commitment to move past the growing pains of implementing a new paradigm and the organizational cultural resistance that may resist this change. However, as the chapters in this volume indicate, over the long-term, the benefits of IBL are worth the effort.

IMPLEMENTATION BENEFITS

A growing body of research suggests that IBL can be an effective teaching and learning strategy if implemented properly with regard to context and the linking of teaching, learning, research, and assessment (Blessinger & Carfora, 2014a, 2014b, 2015; Cuneo et al., 2001, 2012; Justice et al., 2007; Lynch, Kuipers, Pyke, & Szesze, 2005; Vajoczki, Watt, Vine & Xueqing 2011). If utilized appropriately and in a purposeful manner, IBL can produce positive learning effects in the classroom. Since IBL is oriented around authentic and meaningful learning, it is well suited to align better with a student's own value system, learning needs, and life and career aspirations.

One of the key advantages of IBL is that students learn to identify and define meaningful questions and authentic problems to investigate, which they then use inquiry and research to address. This approach is important because it more closely models the way real professionals and scientist learn and create new knowledge. The instructor, as knowledge expert, serves as a resource by guiding the student through the inquiry process, which develops multiple skills in students and fosters collaborative learning since students often work together on projects together. Any inquiry-based approach can work at any level but the instructor, as learning architect, must decide how much guidance is needed and how much direct instruction is needed. Ideally, one chief aim of inquiry-based classrooms is to create a culture of inquiry where the spirit of inquiry is diffused in all learning activities.

IBL does not just involve having students do projects or answer questions posed by instructors during classroom discussions but rather it aims to develop deep discipline-based thinking. IBL is an umbrella term that refers to a strategic approach to learning that often involves, to varying degrees depending on the grade level and discipline and educational context, problem-based learning, project-based learning, and research-based learning. Common inquiry-based elements include: addressing authentic questions/issues, developing higher-order thinking through honing research and communication skills, collaborating with peers and knowledge experts, and developing solutions to complex problems. IBL is a type of active learning strategy because it engages the student in their own knowledge production and epistemic practices of the discipline by putting the main locus of control for learning with the student so that she/he can develop the skills necessary to continue to learn long after leaving the classroom. Thus, in addition to many cognitive benefits, IBL is a more holistic learning strategy for developing important psychological, social, and behavioral qualities and skills necessary for higher-order thinking and lifelong learning.

THEORETICAL FRAMING

In a broad sense, IBL can be viewed as a set of learning and assessment strategies and standards where student learning is grounded in inquiry that is driven by questions and problems relevant to the course and learning objectives (Levy, Lameras, McKinney, & Ford, 2011). IBL is a form of active learning that engages students in a more meaningful, purposeful, and self-regulated way to learn. IBL is applicable and relevant across all disciplines and levels within education and in both formal and informal learning situations (Levy, Little, McKinney, Nibbs, & Wood, 2010). The broad based, ubiquitous applicability of IBL allows its benefits to be used in virtually every aspect of teaching and learning, and also provides a common theoretical framework upon which to design more effective learning environments without the need to create a rigid system of standardization across educational disciplines and levels. The key to effective design is in the contextualization of IBL to the specific level, discipline, course, learning objectives, etc. At the heart of IBL is an inquiry teaching and learning process where all learning activities and assessments are purposefully designed to cultivate knowledge building and higher-order thinking through exploration of authentic and meaningful questions and problems.

The National Research Council (NSES p. 23) defines scientific inquiry as "the diverse ways in which scientists study the natural world and propose explanations based on the evidence derived from their work. Scientific inquiry also refers to the activities through which students develop knowledge and understanding of scientific ideas, as well as an understanding of how scientists study the natural world." This definition reflects how scientists investigate the natural world and how STEM students gain epistemic knowledge and become immersed in the practice of authentic science. Thus, understanding math and science is enhanced when it is grounded in authentic scientific inquiry. At the heart of all inquiry is the ability to identify meaningful problems and to ask meaningful questions that can be answered through scientific inquiry. Inquiry-based science requires students to conduct investigations in order to collect data. It is through the application of the scientific method that students learn to collect and analyze data in order to arrive at plausible evidence-based answers to the questions. Finally, scientific inquiry requires students to communicate and defend their conclusions. Thus, IBL is naturally suited to this scientific process.

IBL has its roots in a constructivist-based educational philosophy – a rich and extensive epistemology whereby learners construct knowledge through interaction with ideas, experiences, and empirical data – and is oriented around three main components:

(1) exploration and investigation (e.g., problem-based learning, collaborative learning, self-directed learning, meaningful learning),
(2) authentic inquiries using contextualized and situated learning (e.g., field learning, service learning, case-based learning), and
(3) research-based approach (e.g., research-based learning, project-based learning, scaffolded learning).

As such, IBL naturally supports different modes of inquiry where the questions to be answered or the problems to be solved are grounded in the relevant epistemological basis (i.e., the type and level of course and discipline, and mode of inquiry). IBL is based on the notion that learning is more effective when one starts with overarching questions and/or problem scenarios. For instance, as in academic and scientific research, one starts with a research question(s) or problem and that drives the entire research process. In like manner, IBL in the STEM fields is driven by an investigation into finding answers to address the questions and/or finding solutions to the problems. In doing so, IBL transforms the teaching and learning process where students also become the teachers to their peers.

A key characteristic of IBL is that it is not prescriptive; rather, it is a broad approach or paradigm (i.e., a set of interconnected theories, concepts, and practices) to teaching and learning. IBL is oriented around process, content, and outcome and is considered a way of learning that focuses on investigating authentic (real-world) questions and problems that are meaningful to learners (Drayton & Falk, 2001). IBL holds that learning is most effective when the learning environment fosters inquiry and enables students to actively engage in knowledge construction in a socially participatory way (e.g., through interaction and collaboration with peers, instructors, and through a variety of relevant resources) (Levy et al., 2011; Biggs, 1996). IBL's conceptual and theoretical roots can be traced back to the works of, among others, Bruner (1961, 1990), Dewey (1910, 1916, 1933, 1938a, 1938b), Freire (1984), Papert and Harel (1991), Piaget (1970, 1971, 1977, 1985), Schwab (1960, 1966), and Vygotsky (1962, 1978).

IBL is a process that builds upon previous knowledge and experience and it leverages multiple ways of knowing. The net effect in this type of learning environment is that students are learning how to learn in that they are learning how the learning process works and are actively engaged in it. They come to realize that the learning process involves properly identifying questions/problems and then applying knowledge and problem-solving skills to collect, analyze and synthesize information and arrive at viable conclusions and solutions. In this way, the learning process mimics the real-world problem-solving process used by scientists and other professions. Thus, the focus is on epistemic knowledge creation and production as well as epistemic practice. The focus is on both theory and practice.

IBL facilitates the production of new knowledge, skills, attitudes, and behaviors through the development of collaborative investigation of questions and problem scenarios using modes of inquiry appropriate to the discipline. This is important because each discipline has evolved within a particular epistemological framework and professional culture. Because IBL is non-prescriptive and respects the epistemological basis of each discipline, the instructor is free to design the course within the established epistemological framework of the discipline and the developmental level of the students while also enjoying the freedom of inquiry to integrate innovative elements into the course to allow the students to be creative in their learning. This is where the pedagogical judgment of the instructor as chief learning architect is important.

Lee (2013) emphasizes the need to balance the challenge and rigor of the course with appropriate support and structure. If the challenge and rigor of the course is well above the grade level of the students, then frustration will likely set in due to cognitive overload. If the challenge and rigor of the course is well below the grade level of the students, then boredom will likely set in. Similarly, if appropriate instructional support and structure is not present, students may go in the wrong direction or pointlessly engage in activities or consume resources that do not yield deep learning. Thus, appropriate instructional design is critical to reaching defined learning outcomes and to match the level of guidance and supervision needed for the particular course. IBL may be unsettling to some students (and instructors) when they are first exposed to it but with the proper support and purposeful course design, they can begin this journey with greater confidence.

CHAPTER OVERVIEWS

In "Inquiry-Based Learning in an Undergraduate Honours Program: Lessons from the Bachelor of Health Sciences Honours Program at McMaster University," by Barb Bloemhof, the author discusses how inquiry-based learning can be used as an approach to create more autonomous, self-regulating lifelong learners. The author examines the demands in this approach placed on both instructors and learners in order to master its principles and integrate it into the teaching and learning process. Specifically, the author examines how the Bachelor of Health Sciences Honours Program at McMaster University in Canada provides an inquiry-based learning model that balances knowledge acquisition and creation with the ability to develop complex transferrable skills, which is a necessary component of lifelong learning.

In "Two Decades of Inquiry-Oriented Learning in First Year Undergraduate Physics Laboratories: An Australian Experience," by Les Kirkup, the author provides a review of the first year physics laboratory program at the University of Technology, Sydney, Australia. The results of the program review showed that student laboratory experiences did not resemble professional practices of physicists, did not give a genuine view of the contribution of physics to everyday life, and did not enhance students' broad capabilities such as communication skills. In light of these findings, the physics department at the University of Technology decided to reform students' laboratory experiences by using an inquiry-based learning approach. This chapter examines the role of the laboratory in undergraduate physics education and the strategies implemented (over a 20-year period) to integrate inquiry-based learning into first year physics laboratory programs.

In "Arnold Arons: Pioneer of Inquiry-Based Learning in Physics," by Richard R. Hake, the author discusses the pioneering efforts of Arnold Arons, one of the founding fathers of U.S. Physics Education Research, who made pioneering efforts in inquiry-based methods of science education. The author discusses how the instructional methods advocated by Arons emerged from his half century effort to improve teaching and learning in introductory science courses. Arons' approach involved the use of Socratic questioning to probe students understanding of the course content and as a means to foster students' reasoning and scientific way of thinking. Arons emphasized 11 specific principles to fostering improved teaching and learning in introductory science education.

In "MIT's Introduction to Biological Engineering: A Longitudinal Study of a Freshman Inquiry-Based Class," by Natalie Kuldell and Rudolph Mitchell, the author describes how a project-based and mentoring inquiry approach is used in an introductory biological engineering course at MIT to create a more effective teaching and learning environment. This approach involves forming small teams of freshman students and pairing them with senior MIT student in order to design a biotechnology that addresses a real-world task of their choosing. The authors discuss a multi-year study of the class experience which utilized post-surveys and semi-structured interviews of two freshman cohorts and a retrospective survey of three freshman cohorts. The findings of the research study suggest that students perceive that academic gains were achieved. Specifically, students reported that they gained greater understanding of the course content and that they are better able to make more meaningful connections between scientific ideas they learned, their personal goals, and their vision of their future selves.

In "Problem Based Learning Approaches to Teaching Chemistry," by Dylan Powell Williams, the author examines how problem-based learning supports a first-year chemistry program at the University of Leicester, United Kingdom. The author provides an overview of this approach to teaching and learning in chemistry and the author examines the challenges encountered. The author also examines how they adapted problem-based learning in response to student and staff feedback.

In "An Intentional and Facilitative Student-Centered Approach to Enhance Inquiry-Based Learning Utilizing Authentic, Hypothesis-Driven Research," by Amelia J. Ahern-Rindell, the authors discuss how her genetics laboratory course allows students to use hypothesis-driven research via small groups to foster authentic learning experiences. The author explains how the course emphasizes the scientific process and the practical application of science. The author explains how scaffolding is used to improve critical thinking, communication, and data interpretation, as well as how the guided-discovery approach is used to develop creativity by posing questions that form the basis of their hypothesis-driven work.

In "IBL and Mathematics: The Necessity of Flexible Knowledge," by Vanessa R. Pitts Bannister and Gina J. Mariano, the authors discuss cases involving pre-service mathematics teachers. The authors discuss cases that illustrate teachers' understanding of perspectives as they analyze student work involving tasks that call for translations between algebraic and graphical representations. The authors discuss findings that suggest how flexible or inflexible knowledge structures influence participant's attempt to analyze student thinking. The authors also discuss findings that stress the importance of flexible teacher knowledge in their ability to identify student thinking perspectives.

In "Advantages of Science Education Outdoors through IBSE Methodology," by Ana Cristina Tavares, Susana Silva, and Teresa Bettencourt, the authors discuss the program Science Education Outdoors is an active learning program that provides students with direct experience with natural phenomena. The authors discuss three case studies based on training course provided at the Botanical Garden of the University of Coimbra in Portugal. The cases utilize inquiry-based science education to promote more effective and meaningful teaching and learning in science. The findings of the study revealed that participants examined in the cases considered IBSE to be an effective approach to teaching and learning and professional development.

In "Inquiry Based Learning in Action: Reflections on an Interdisciplinary Master's Program in the Health and Life Sciences," by Marjolein B. M. Zweekhorst, Wanda S. Konijn, Jacqueline E. W. Broerse and Jeroen Maas, the authors reflect on the need to address complex social problems through collaboration among stakeholders from science and society and the integration of knowledge from multiple disciplines. In this chapter, the authors suggest that students can be trained in interdisciplinary research by using a gradual approach that moves students from structured inquiry to open inquiry. The role of the instructor is crucial in facilitating this process.

In "A Narrative-Based Bachelor of Science in Nursing Program: Experiences and Insights," by Baiba Zarins and Lorraine Carter, the authors examine a program called the Scholar Practitioner Program, which is oriented around narrative inquiry, cognitive apprenticeship, and experiential learning with community-based health science centers. The aim of the program is to build on students' prior university learning. The program fosters a spirit of inquiry and research culture that cultivates self-directed learning and learner responsibility. By immersing the students in professional practice in real-world environments, students are better able to apply knowledge gained from formal learning with meaningful professional experience.

In "Modeling Scientific Inquiry to Guide Students in the Practices of Science: The Ed3u Teaching Model of Conceptual Change in Action," by Richard E. Shope and William F. McComas, the authors discuss how a program called EcoVoices is used as a vehicle for K-12 learners to practice science by engaging in inquiry of urban environmental science. The authors discuss the principles behind the ED3U instructional model and how it is applied in this case. The authors discuss how this program provides students with an opportunity to practice science in an authentic setting using an inquiry-based learning approach.

In "Utilising Context in Learning," by Wendy Trimmer and Philip Hawes, the authors discuss the implementation of an inquiry-based approach to teaching and learning at Whitireia Community Polytechnic (Whitireia), a post-secondary educational institution known for its culturally diverse student body. The authors discuss how context-based learning has been used across a range of courses in several health disciplines. Context-based learning encompasses a simulated learning environment and a purpose built physical environment.

In "Evaluating European Inquiry-Based Education Projects through The Lens of Evaluation Capacity Building," by Fabio Dovigo and Vincenza Rocco, the authors explore evaluation practices of inquiry-based science education. The authors analyzed the European project INQUIRE within the framework of Evaluation Capacity Building theory, which aims to develop rigorous evaluation capabilities. The authors analyzed evaluations used within courses held in Italy over a two-year period. Their findings suggest that the INQUIRE project has helped practitioners to improve their ability to conduct effective evaluations. The authors also explore some concerns and make some recommendations based on their conclusions.

In "Inquiry-Based Learning in Mathematics," by Mindy Capaldi, the author describes the challenges faced by mathematicians to make math learning more enjoyable and to make the teaching and learning of math more effective and meaningful. The author examines how inquiry-based learning can help address this challenge. The author examines how inquiry-based learning teachers do not lay out all of the formulas and theorems as previous knowledge but rather focus on the creative process of solving mathematical problems, and in so doing, help make the learning of math more enjoyable for both teachers and students as well as simulating the real experiences of mathematicians.

In "Enhancing Inquiry-Based Learning Environments with The Power of Problem-Based Learning to Teach 21st Century Learning and Skills," by Henk Huijser, Megan Y. C. A. Kek, and Ruth Terwijn, the authors discuss how problem-based learning can be adapted to enhance inquiry-based learning environments. The authors provide a case study of a first-year nursing course in an Australian University. The authors argue that a carefully defined set of learning outcomes together with a properly designed set of inquiry-based learning activities provide a more flexible learning environment in which learners are better able to acquire 21st century knowledge and skills.

CONCLUSION

In IBL, students actively seek solutions to problems or questions. The level and degree of guidance and direction received from the instructor is dependent on the context in which the course is conducted. Some courses may require a lot of guidance and other courses may require less. The instructor, as instructional leader and learning architect, decides what type of inquiry is most appropriate. In an IBL classroom, students are not just sitting there waiting for the instructor to provide them with all the answers and the instructor is not just spoon-feeding students a steady diet of information that students could readily find in the textbook for themselves. Rather than being passive learners, students learn to become more active learners.

In IBL, the focus moves from what the instructor is teaching to what the student is learning. Thus, IBL is not so much a specific instructional method as it is an active learning strategy and approach that aligns learning with how learning works in real-world, professional settings. In IBL, the instructor moves from being a transmitter of information where information is spoon-fed to students with a daily diet of lectures to an expert guide who facilitates the learning process and cultivates an environment where the students are required to take increasing responsibility for their own learning. In short, the aim of IBL is to transform students from instructor-dependent direction-followers to more self-regulating problem-solvers.

In this volume, a range of case studies and other empirical research have been presented on how IBL is being used across a range of STEM courses. These findings, together with the current research literature on IBL, strongly suggest that the IBL approach has great potential to enhance and transform teaching and learning. However, this volume also provided a discussion of the challenges that may arise when implementing IBL. As the chapters in this volume illustrate, IBL is an approach and general framework providing educators with new possibilities to enhance and transform how they teach and interact with students and how students interact with each other and with the course content. IBL helps expand our notions about what it means to teach and learn in the modern era.

REFERENCES

Barrow, L. (2006). A brief history of inquiry: From Dewey to standards. *Journal of Science Teacher Education*, *17*, 265–278.

Bell, T., Urhahne, D., Schanze, S., & Ploetzner, R. (2010). Collaborative inquiry learning: Models, tools and challenges. *International Journal of Science Education*, *32*(3), 349–377.

Bereiter, C., & Scardamalia, M. (2006). Education for the knowledge age: Design-centered models of teaching and instruction. In P. A. Alexander & P. H. Winne (Eds.), *Handbook of educational psychology* (2nd ed., pp. 695–713). Mahwah, NJ: Erlbaum.

Biggs, J. (1996). Enhancing teaching through constructive alignment. *Higher Education*, *32*, 347–364.

Blessinger, P., & Anchan, J. P. (2015). *Democratizing higher education: International comparative perspectives*. New York, NY: Routledge Publishing.

Blessinger, P., & Carfora, J. M. (2014a). *Inquiry-based learning for faculty and institutional development: A conceptual and practical resource for educators*. Bingley, UK: Emerald Group Publishing Limited.

Blessinger, P., & Carfora, J. M. (2014b). *Inquiry-based learning for the arts, humanities, and social sciences: A conceptual and practical resource for educators*. Bingley, UK: Emerald Group Publishing Limited.

Blessinger, P., & Carfora, J. M. (2015). *Inquiry-based learning for multidisciplinary programs: A conceptual and practical resource for educators*. Bingley, UK: Emerald Group Publishing Limited.

Bransford, J. D., Brown, A. L., & Cocking, R. (2000). *How people learn*. Washington, DC: National Academy Press.

Bruner, J. S. (1961). The act of discovery. *Harvard Educational Review*, *31*(1), 21–32.

Bruner, J. S. (1990). *Acts of meaning*. Cambridge, MA: Harvard University Press.

Carfora, J. M. (2011). Navigating between teaching, learning and inquiry. *International HETL Review*, *1*, 3–8. Retrieved from https://www.hetl.org/opinion-articles/teaching-learning-inquiry

Cuneo, C., Harnish, D., Roy, D., & Vajoczki, S. (2012). Lessons learned: The McMaster inquiry story from innovation to institutionalization. In V. Lee (Ed.), *Inquiry-guided learning. New directions in teaching and learning* (Vol. 129, pp. 93–104). New York, NY: Wiley.

Cuneo, C., Inglis, S., Justice, C., Lee, W., Miller, S., Rice, J., … Warry, W. (2001). Thinking and doing outside the box: Interdisciplinary inquiry learning partnerships. *Research and Development in Higher Education, 24*, 15–22.

Daloz Parks, S. (2000). *Big questions, worthy dreams: Mentoring young adults in their search for meaning, purpose and faith*. San Francisco, CA: Jossey-Bass.

Dewey, J. (1910). Science as a subject matter and as method. *Science, 31*(787), 121–127.

Dewey, J. (1916). Methods in science teaching. *The Science Quarterly, 1*, 3–9.

Dewey, J. (1933). *How we think. A restatement of the relation of reflective thinking to the educative process* (Rev. ed.). Boston, MA: D. C. Heath.

Dewey, J. (1938a). *Experience and education*. New York, NY: Collier Books.

Dewey, J. (1938b). *Logic: The theory of inquiry*. New York, NY: Holt, Rinehart and Winston.

Doherty, A., Riordan, T., & Roth, J. (2003). *Student learning: A central focus for institutions of higher education*. Milwaukee, WI: Alverno College Institute.

Drayton, B., & Falk, J. (2001). Tell-tale signs of the inquiry-oriented classroom. *NASSP Bulletin, 85*(623), 24–34.

Eslinger, E., White, B., & Frederiksen, J. (2008). Supporting inquiry processes with an interactive learning environment: Inquiry Island. *Journal of Science Education and Technology, 17*, 610–617.

Freire, P. (1984). *Pedagogy of the oppressed*. New York, NY: Continuum Publishing Company.

Garrison, D. R., Anderson, T., & Archer, W. (2000). Critical inquiry in a text-based environment: Computer conferencing in higher education. *The Internet and Higher Education, 2*(2–3), 87–105.

Gredler, M. (2009). *Learning and instruction: Theory in practice*. Upper Saddle River, NJ: Merrill Pearson.

Hmelo-Silver, C., Duncan, R., & Chinn, C. (2007). Scaffolding and achievement in problem-based and inquiry-learning: A response to Kirschner, Sweller and Clark (2006). *Educational Psychologist, 42*(2), 99–107. doi:10.1080/00461520701263368

Justice, C., Rice, J., Warry, W., Inglis, S., Miller, S., & Sammon, S. (2007). Inquiry in higher education: Reflections and directions on course design and teaching methods. *Innovative Higher Education, 31*(4), 201–214.

Kirschner, P., Sweller, J., & Clark, R. (2006). Why minimal guidance during instruction does not work: An analysis of the failure of constructiviest, discovery, problem-based, experiential and inquiry-based teaching. *Educational Psychologist, 41*(2), 75–86. doi:10.1207/s15326985ep4102_1

Kovbasyuk, O., & Blessinger, P. (2013). *Meaning-centered education: International perspectives and explorations in higher education*. New York, NY: Routledge.

Kuh, G. (2008). *High impact educational practices: What they are, who has access to them, and why they matter*. Washington, DC: AAC&U.

Lee, V. S. (2013). Supporting students' search for a meaningful life through inquiry-guided learning. In O. Kovbasyuk & P. Blessinger (Eds.), *Meaning-centered education, international perspectives and explorations in higher education*. New York, NY: Routledge.

Lee, V. S., Greene, D. B., Odom, J., Schechter, E., & Slatta, R. W. (2004). What is inquiry-guided learning? In V. S. Lee (Ed.), *Teaching and learning through inquiry: A guidebook for institutions and instructors* (pp. 2–16). Sterling, VA: Stylus.

Levy, P., Lameras, P., McKinney, P., & Ford, N. (2011). The features of inquiry learning: theory, research and practice. Pathway to Inquiry Based Science Teaching, (Deliverable 2.1). European Commission: CSA-SA Support Actions, Project Number 266624. Retrieved from http://www.pathwayuk.org.uk/what-is-ibse.html

Levy, P., Little, S., McKinney, P., Nibbs, A., & Wood, J. (2010). *The Sheffield companion to inquiry-based learning*. Sheffield: Centre for Inquiry-Based Learning in the Arts and Social Sciences, The University of Sheffield.

Lynch, S., Kuipers, J., Pyke, C., & Szesze, M. (2005). Examining the effects of a highly rated science curriculum unit on diverse students: Results from a planning grant. *Journal of Research in Science Teaching*, 42, 921–946.

Mezirow, J. (1991). *Transformative dimensions of adult education*. San Francisco, CA: Jossey Bass.

National Research Council. (1996). *National science education standards*. Washington, DC: National Academy Press.

Papert, S., & Harel, I. (Eds.). (1991). Constructionism: Research reports and essays 1985–1990. *The Epistemology and Learning Research Group, the Media Lab, Massachusetts Institute of Technology*. Norwood, NJ: Ablex Publishing Corporation.

Piaget, J. (1970). *Science of education and the psychology of the child*. New York, NY: Orion Press.

Piaget, J. (1971). *Biology and knowledge*. Chicago, IL: University of Chicago Press.

Piaget, J. (1977). *Intellectual evolution from adolescence to adulthood*. Cambridge, MA: Cambridge University Press.

Piaget, J. (1985). *The equilibration of cognitive structures: The central problem of intellectual development*. Chicago, IL: University of Chicago Press.

Sandoval, W. A., & Reiser, B. J. (2004). Explanation-driven inquiry: Integrating conceptual and epistemic supports for science inquiry. *Science Education*, *88*, 345–372.

Schwab, J. (1960). Inquiry, the science teacher, and the educator. *The School Review*, *68*(2), 176–195.

Schwab, J. (1966). *The teaching of science*. Cambridge, MA: Harvard University Press.

Spronken-Smith, R. (2012). Experiencing the process of knowledge creation: The nature and use of inquiry-based learning in higher education. Paper prepared for International Colloquium on Practices for Academic Inquiry. Retrieved from https://akoaotearoa.ac.nz/sites/default/files/u14/IBL%20-%20Report%20-%20Appendix%20A%20-%20Review.pdf. Accessed on May 14, 2014.

Vajoczki, S., Watt, S., Vine, M. M., & Xueqing, L. (2011). Inquiry learning: Level, discipline, class size, what matters? *International Journal for the Scholarship of Teaching and Learning*, *5*(1), 1–11.

Vygotsky, L. S. (1962). *Thought and language*. Cambridge, MA: MIT Press.

Vygotsky, L. S. (1978). *Mind in society*. Cambridge, MA: Harvard University Press.

INQUIRY-BASED LEARNING IN AN UNDERGRADUATE HONOURS PROGRAM: LESSONS FROM THE BACHELOR OF HEALTH SCIENCES HONOURS PROGRAM AT McMASTER UNIVERSITY

Barb Bloemhof

ABSTRACT

Inquiry-based learning is a fruitful way to create "effective, independent learners" (Baird, 1988, p. 142) and set up the pattern for lifelong learning, but inquiry skills demand significant practice to master and incorporate in both academic and personal dimensions. The Bachelor of Health Sciences Honours program at McMaster University provides a model of an undergraduate program that balances knowledge and the complex transferrable skills associated with inquiry. By devoting considerable resources to the first year experience and integrating the curriculum so that meaningful use is made of the inquiry skills developed there, the program fosters the curiosity, confidence, and capability of students.

Inquiry-Based Learning for Science, Technology, Engineering, and Math (STEM) Programs: A Conceptual and Practical Resource for Educators
Innovations in Higher Education Teaching and Learning, Volume 4, 21–40

ISSN: 2055-3641/doi:10.1108/S2055-364120150000004002

The curriculum demonstrably meets or exceeds the standards for quality set out by governing bodies within and outside of the university known as the degree level expectations. The current chapter provides an overview of the program, including lessons for anyone engaged in curriculum design that builds undergraduate research capability.

McMaster University has been using inquiry approaches to instruction since 1980; the Arts and Sciences program is built around inquiry, and individual faculties at the university featured inquiry courses to greater or lesser extent (Roy, 2007). The focus of the current chapter is on the Bachelor of Health Sciences Honours undergraduate degree program at McMaster University (BHSc), designed to build research capacity by using inquiry in an integrated curriculum of student self-direction and authentic learning experiences over the course of a four-year program of study.

Exceptional ideas are often grounded in a personal experience. For Del Harnish, the germ of an exceptional idea came to him when he was asked by a colleague to substitute-teach a fourth year course in virology.[1] Dr. Harnish gave the students a well-organized and synthesized talk integrating about six months of engaging scholarly literature in the space of 50 minutes. The students balked: they wanted a standard lecture, they were not curious about the highly interesting things that were contained in that literature despite the fact that they had been studying biochemistry by choice all through their degree. The experience started a series of questions to self: what explains the absence of an intrinsic curiosity about the material a student is majoring in?; in a field where information has a half-life under five years, how could students near the end of their studies have avoided reading any peer-reviewed scholarly papers all the way through?; what sort of program and curriculum design could repair the preference for lectures and assessments built around "tell us what we need to know so we can give it back to you on the exam"?

The answers to these questions took about ten years to craft, but their manifestation in action was path-breaking. The Bachelor of Health Sciences Honours program began in 2000 with a first-year intake of around 80 students, growing to 200 students currently for a steady state program enrollment of about 800 students. The program offers a place for an undergraduate to study a wide range of health sciences and health policy using inquiry, problem-based learning, laboratory work, small group tutorials and lectures. Students build facility with critical thinking in each year of study. However, the strategic use of inquiry in the program is crucial to educating for capability, because learning environment dominates curriculum in effecting that goal (Harnish, 2012, p. 107). Inquiry classes are capped at 20 students in the first and third year courses; the emphasis is on process, and not predetermined content, with students making autonomous defensible decisions about their learning. In fact, even the first-year inquiry knowledge theme is codetermined collaboratively by students, although of course the skills foundation is common across the year. In all program classes, students spend time in activities that use the inquiry skills discussed below, with enormous amounts of individualized formative feedback and assessments that only become summative after the student has had opportunities to practice important skills. The approach supports facility with self-direction in parallel with skills development and personal growth.

The reputation of the BHSc program has spread quickly over just ten graduating classes. With over 3,500 applications for less than 200 intake spots in 2014, it is arguably the most competitive program of its kind in Canada. BHSc graduates would be valuable "gets" across the allied health professions tertiary programs, and in MBA and graduate programs that receive them, because they are robust independent thinkers equipped to soar in the presence of academic challenge and ambiguity. As undergraduates, BHSc students are happily taken on by faculty across the university to work with supervisors on their undergraduate theses, or on community or academic research projects, anywhere a critical and resourceful mind is needed.[2]

Dr. Harnish (2012, p. 108) is right when he says that "[b]right young people will succeed in most environments no matter what we do," but that is not the whole story. A commitment to authentic learning experiences that use inquiry as a vehicle for developing crucial transferrable skills is critical to fostering undergraduate research capacity. In fact, the program reads like the recommendations list of Boyer's eponymous commission report (Kenny, 1998). The focus on collaboration and openness to new information are key drivers in how the program's students become

> effective, independent learners – a person who is both willing and able to take responsibility for, and control over one's own learning. Such a person can make informed, purposeful decisions about planning, managing and evaluating personal learning; he or she can thereby function productively in and contribute to the betterment of society at large. (Baird, 1988, p. 142)

This chapter explores the structure of and rationale for a degree program built on inquiry, and provides some insights for those who want to build inquiry-based programs.

THE GOALS OF INQUIRY

Inquiry is "*a self-directed, question-driven search for understanding*" (Hudspith & Jenkins, 2001, p. 9, italics in original). Inquiry orients the course toward the process of learning. Candy (1991) characterizes four dimensions of self-direction: personal autonomy, self-management in learning, independence, and learner control of the instruction. By giving them tangible control over many aspects of the curriculum, inquiry changes students' construction of their role in the classroom, and creates a new relationship between the student and their learning (Postman & Weingartner, 1969, p. 27). If the instructor is successful in credibly communicating the freedom to self-direct, then interest and curiosity provide much motivation for the student's intellectual development.

One fundamental goal of inquiry is to foster the skill of following a question that is important and currently unknown to the learner (Faculty of Health Sciences, 2014; Justice et al., 2007, p. 202; Lee, 2007). Removing from the instructor the responsibility to justify why it is important to ask this or that question kindles in the learner a process of exploration into productive habits of mind and valuable transferable skills (Bloemhof, 2007, p. 32; Justice et al., 2001, p. 3; Knowles, 1975, p. 23; Reynolds, Saxon, & Benmore, 2006, p. 359; Tough, 1971, p. 40). During inquiry, students must also learn how to set priorities and manage their time, how to orient their learning gaps around a central researchable question, how to find and evaluate scholarly resources, how to anticipate possible answers, how to integrate findings and insights into a new product that goes beyond a summary of others' work, and how to communicate that new product appropriately. A number of self-management skills, such as working with peers and experts, self-assessing strengths and weaknesses, and critical appraisal of available methods and findings, are part of the process of inquiry.

A typical course of inquiry predictably engenders proficiencies that are valuable in a complex and changing world, and critical thinking is just the most obvious of these. Figuring out what one needs to learn and examining assumptions applies broadly in life. The student starts with a vague topic area and begins to research around it, to learn important background knowledge that informs a first question, typically a "why" or "how" question. Peer and instructor feedback on the student's initial research question gives rise to an examination of the assumptions made; the student will typically research well beyond what they eventually report, honing in on the most lively or important elements of the issue, and refining the question a number of times as they incorporate what they have learned. Proficiency at accessing and using the scholarly and grey literatures, subject matter experts and their own good minds all come into play in sleuthing a defensible new belief or understanding about a key puzzle related to the topic area. As well, the student will need to devote effort to self-management skills around staying organized, managing time and knowing when to wrap up or even abandon a particular line of research. Add to these complex skills an array of listening and questioning skills that the student will use in order to respond appropriately to peer and instructor feedback.

The production of some sort of student-chosen deliverable involves another vector of proficiencies around collaboration, rhetorical effectiveness and information integration, as well as considering what is worth delivering. Typically, students in a first or second course of inquiry will be working in groups, which improve learning outcomes and motivate students (Slavin, 1996; Springer, Stanne, & Donovan, 1999), so collaborative skills around taking ownership, committing to actions and following through become prominent. Class time will often be an opportunity for discussion, with the attendant democracy of providing a place for a number of perspectives, where reasoned argument together with careful listening can be exercised (see Brookfield & Preskill, 1999).

Inquiry instructors also dedicate considerable time to honing complex skills. Facilitating inquiry may involve a change in the instructor's paradigm (Knowles, 1975, p. 31) and ongoing flexibility and responsiveness (Roy, 2007, p. 41; Trim, Harnish, Lovric, & Roy, 2007, p. 61). Academics with humility about the limits of their knowledge and enthusiasm for discovery make the best inquiry instructors in part because they model valuable learner attitudes and create a safe place for the learner to take risks. In fact, the truth that we are all learners is a useful antidote to the hegemony of the typical post-secondary classroom. The inquiry instructors in the BHSc program demonstrate a continuing commitment to questioning, self-assessment, and evidence-based change with respect to their teaching. The next section describes the program that has evolved as a result.

STRUCTURE OF THE CURRICULUM

The Bachelor of Health Sciences Honours Program is designed to foster students' proficiencies with collaboration and self-directed learning. Staff begin creating the learning environment in the summer before classes start. First, the letter offering admission to the program makes reference to the value of *relevant* information in making a decision about where to study, and directs the candidate to three processes they might use to gather that information. This approach is quite different from persuading with selected facts; perhaps as a result, proportionately more offers are taken up by students in the BHSc program than in a number of ivy-league universities including Harvard.[3] An air of genuine warm welcome attends interactions at the mid-summer freshman barbeque and at "first week" activities in September.

Once classes start, students begin to perceive a potent expectation for their success as they hear about the goals and objectives of the program, and about the methods by which students will achieve them. Inquiry is defined using at least three different conceptual frameworks: the "crap detecting" framework of Postman and Weingartner (1969), the "Inquire, Research/Reflect, Evaluate, Construct" (IREC) model (Faculty of Health Sciences, 2014), and the instructor's own discursive treatment of the program's list of inquiry skills.[4] Metacognition is both modeled and invited, so students see what and how they are learning right from the start. The course expectations arc transparent and accessible, though it is not clear how credible this transparency initially is. In their program courses, students are told clearly and in diverse ways what to expect in words that inspire confidence and aspiration rather than oppressive language that suggests burden and struggle. For example, in a number of first year syllabi students read that their role is to be eager and curious and invited to have fun in the course. Finally, as Table 1 shows, the program deliberately incorporates constructive alignment (Biggs, 1996) by sequencing the courses to support the co-curricular goals both within each year level and progressively as students transition from one level to another.

In this positive learning environment the faculty are candid about their own scholarly curiosity and skepticism when discussing any scholarship or ideas, which induces a norm of respect for evidence-based knowledge and its creation. A wide range of instructional methods are used at carefully considered junctures, but the backbone of the program is self-directed learning, and the dominant form of self-directed learning in the program

Table 1. Bachelor of Health Sciences (Honours) Program Curriculum.

Name of Course (Course Host Department)[a]		
Year 1: 7 half-year equivalent courses plus 3 open electives	Elective	Elective
	Psychobiology (BHSc)	Elective
	Chemistry I (CHEM)	Chemistry II (CHEM)
	Cellular and Molecular Biology (BHSc)[d]	
	Level 1 Inquiry (BHSc)[d]	
Year 2: 7 half-year equivalent courses plus 3 open electives	Elective	Elective
	Health, Attitude, and Behaviour[b] (BHSc)	Elective
	Epidemiology (BHSc)	Statistics (BHSc)
	Human Physiology and Anatomy I (BHSc)	Human Physiology and Anatomy II (BHSc)
	Level II Inquiry (BHSc)	Cell Biology Inquiry (BHSc)
Year 3: 4 half-year equivalent courses plus 3 open electives	Elective	Elective
	Elective	Elective
	Elective	Elective
	Critical Appraisal of the Medical Literature (BHSc)	Health Systems and Health Policy (BHSc)
	Level III Inquiry[b] (BHSc)	Inquiry Project[b] (BHSc)
Year 4: 3 (or 4 in thesis option) half-year equivalent courses plus 7 (6) open electives	Elective	Elective
	Elective	Elective
	Elective	Elective
	Collaboration and Peer Tutoring[c] (BHSc)	Elective in project option; curricular space for thesis
	Project (full year) or Thesis (1.5 × full year weight) (BHSc)	

Notes: Level I students also complete a workplace hazardous materials information safety course. Level II students also complete a biosafety training course.

[a]BHSc = Bachelor of Health Sciences (Honours); CHEM = Chemistry.

[b]Can be taken in fall or winter term.

[c]Collaboration and Peer Tutoring occurs over all four years, although registration for the course occurs as one half-course equivalent credit weight in year 4.

[d]Full year course; equivalent to two half-year courses.

is inquiry. Like all self-directed learning, in inquiry students have considerable autonomy to decide what is worth learning, with a concomitant expectation that they will also justify that choice using level-appropriate evidences. This is most challenging in first year (Level I), when the roughly 200 new students are split into groups of no more than 20 per class section for a full year working with the skills that are fundamental to an effective learner and collaborator.[5] Level I Inquiry is assessed by means of individual interviews with the facilitator to collegially discuss the evidence for skills development. The common course philosophy is organized around measurable learning goals listed in Table 2 that are used and further developed in other BHSc required and elective courses across all year levels. In particular, two other introductory courses (Cellular and Molecular Biology, and Psychobiology) are designed to require that a student continuously *use* the skills of inquiry, in the manner described in Biggs (1996, p. 309, quoting Shuell [1986]). In other words, the curriculum is integrated to deliberately provide opportunities in all concurrent courses to practice effective and collaborative learning. Students quickly see the relevance of their inquiry skills in other required program courses.

In second year, all students take an one-term collaboratively taught Level II Inquiry. In random fixed groups of seven or eight students, they spend the term developing further proficiency with the inquiry skills through exploring an authentic problem relevant to human diseases. Each group meets for 40 minutes every other week with their instructor to present their work in progress and to receive feedback. Students are assessed on collaborative project writing and a group presentation, but half of their grade is determined on an individual basis by their skills proficiency during these biweekly check-in meetings. There is another required second-year course, again taught using an inquiry approach, in which students work collaboratively in new randomly assigned fixed teams on applied problems in cell structure and function. Aided by peer tutors to facilitate the group's process, activities foster awareness of the current boundaries and the limits of knowledge. Assessment here is again equally split between group and individual activities.

Having fostered self-management, collaboration and information literacy skills, the upper years of the program provide opportunities to research individually, with far fewer required courses and many more electives. The required Level III Inquiry course is again limited to no more than 20 students per class section, but this time individual work may begin to dominate group work. Instructors identify a theme that students self-select into by registering according to their interests and utilizing information

Table 2. Essential Skills Objectives: Level I Inquiry.

Skill	Description
Personal awareness	The ability to understand yourself and how your behavior impacts others • Self-organization (achieving personal effectiveness, being emotionally present, achieving balance, determining a sense of self, setting priorities, staying on track and managing time)
Problem identification	The ability to ask and refine questions
Problem solving	The ability to determine what needs to be learned in order to answer questions, identify appropriate resources for learning, and use them effectively • Information literacy skills (being conscious of the research process as it takes place) • Identifying sources of information (in the library collection, on the web, from experts, etc.) • Evaluating information content and context • Using information appropriately to answer a question • Reflecting on and evaluating the research process • Constructing and deconstructing knowledge
Professional communication	The ability to communicate effectively and appropriately with others • Verbal, nonverbal and written • Peers, faculty, staff, parents, community members, etc.
Peer collaboration	The ability to work effectively with others • Working with another person and a group • Identifying individual and group strengths and weaknesses • Dividing responsibility • Following through • Teaching each other and learning from each other • Giving and receiving constructive feedback • Dealing with conflict
Personal/Peer evaluation	The ability to evaluate strengths and weaknesses of self and others (formally, informally, and often)

Source: BHSc 1E06 Level I Inquiry 2014 course syllabus; Faculty of Health Sciences (2014).

gleaned from upper year students about their experience of the themes. The goals of inquiry do not change, however (third year) students are expected to execute these process skills at an appropriate level of sophistication and refinement. In this way, Level III Inquiry can be a bridge between the group work of the earlier years and the autonomy of the senior project or thesis in fourth year. Some Level III Inquiry instructors give the students the responsibility for defining and setting the course activities and assessments for all students, based on consensus, in a collaborative discussion process. Another required third-year "course" is the individual inquiry project in which students individually define a project, identify a supervisor to work with, negotiate their goals and schedules, and progress independently with periodic meetings for feedback and guidance. Similarly, many of the third-year elective courses provide opportunities for substantive practice with self-direction and goal setting.

Although comprised of a wide range of elective courses, Level IV has only two required elements: either a senior thesis or project, at the student's discretion; and a peer tutoring experience. The senior thesis or project is just individual inquiry, completely self-directed with the feedback of a faculty supervisor that the student approaches with a proposed research question. The peer tutoring experience has a number of outcomes, one of which puts these experienced inquirers back in the Level I and Level II inquiry classrooms to help foster self-direction in the younger students. Many senior students value the rich integration of their own inquiry skills as a peer tutor to novices adjusting to inquiry-based learning; indeed the course was originally proposed by BHSc students themselves.

Analysis of the Program's Structure

The next task is to determine what this curriculum structure provides to the learner. In the early 2000s, Ontario joined other jurisdictions in thinking about quality in the post-secondary education sector (Council of Ontario Universities, 2011, p. 14). Degree level expectations are generic statements of learning outcomes that have a consistent resonance in any degree program regardless of discipline, spanning depth and breadth of knowledge, knowledge of methodologies, capacity for research and scholarship, application of knowledge, communication skills, awareness of the limits of knowledge and autonomy and professional capacity (Council of Ontario Universities, 2011, p. 18). Inquiry-based learning in the Bachelor of Health Sciences Honours program is an effective means of meeting degree level expectations.

Fully one-fifth of the first-year course time is spent in a supportive, democratic, and respectful inquiry-based learning environment, with no more than 20 students in a class working on defining and refining learning goals and exploring the dynamic array of resources available to address those goals. However, the world needs resourceful, team-oriented, self-directed lifelong learners who are tolerant of change and who can cope effectively with ambiguity. Keeping first- and third-year inquiry classes small makes the codetermination of learning goals feasible, and fosters progressive comfort with the autonomy that drives self-direction. Small learning groups in the more knowledge-based first-year required courses and in the content-rich Level II Inquiry course provides ample support for developing proficiency with the collaborative and self-management skills that serve the learner in the upper years and beyond the degree. Thus, the program builds in redundant practice with the most important transferrable skills: autonomy, collaboration, and resourcefulness in the face of ambiguity. Information literacy, knowledge mobilization, and research capability can be layered on and developed quite quickly with these foundational skills in place. The evidence gathered by Trim (2006, p. 55) suggests that it is most effective to jump right into these latter research proficiencies when students enter the program, before they confound information literacy with "going to the library."

Indeed, it is inappropriate to expect students to take charge of their learning without having adequate curricular space devoted to developing the skill of taking and capitalizing on intellectual freedom. Becoming a proficient autonomous learner involves a complex set of skills that can take a long time to develop (Trim, 2006, p. 167). Supporting students to define learning goals and choose strategies to meet them, and helping them learn to self-assess their proficiency as learners, is just what Knowles (1975, pp. 34–37) advises (see also Tough, 1971). Once students have developed as autonomous and self-directed learners, it would be inappropriate to tell them what to do for their senior capstone experiences (for they may elect to have several). Every upper-year student has at least one experience of arranging, proposing, and executing independent research. Students build on the breadth of knowledge they gain in the first and second years, begin to fledge as researchers in their third year, and thus should be fully research ready for their fourth year senior project or thesis.

To extend the open conversations between students, staff, and faculty, the program uses classroom management software that embodies transparent collaboration for all curricular and extracurricular activities. This software lets users send and unsend messages, view which recipients have read the message and when, and flexibly create subspaces in which computer-mediated collaboration can take place. Students are encouraged to organize themselves within the software in whatever ways make that simple, including in students-only spaces that are private even from the administration. Easy collaboration also justifies the recommended note-taking software; as well, students are encouraged to use the most popular originality-checking software *proactively* on their own essays *before* they hand them in, rather than at their instructor's request without the benefit of seeing the originality report.

There is an air of collegiality in the way students in the program are spoken to, reinforced in the practice of expecting full participation in midterm (formative) and final (summative) course evaluations, and in the credence given to student input into program change decisions. Students are encouraged to practice using the skills that they have been empowered with, repeatedly in their courses and also in their contributions to curricular and social aspects of university life. In this way, the program is designed to build the students' facility, confidence and credibility in taking ownership of their learning.

It is hard to believe that either students or teachers could truly be satisfied with the traditional lecture course. It is unsatisfying as a student to come out of a course with little useful or durable knowledge, and it is tedious as an instructor to be expected to provide students with their thoughts on any particular issue. As autonomous learners in an inquiry format, students learn about complexity, creativity, and the art of questioning as a way "to develop the skills needed to bring research to bear on the understanding of a central question" (Hudspith & Jenkins, 2001, p. 10). Inquiry certainly is an alternative to simple transmission of learning (Kenny, 1998, p. 12). More than this, an integrated collaborative course of study like the BHSc inquiry-based learning program has the potential to inspire and transform the learner.

ASSESSING THE EFFICACY OF THE BACHELOR OF HEALTH SCIENCES HONOURS PROGRAM

The priorities of an academic program are clearest from how the learning actually happens, not from what the brochure says. At the time of writing this chapter, the BHSc program was in the process of a second external review, with results expected in mid-2015. In the first external program review, there was "ample evidence [of] breadth and depth of knowledge and methods of inquiry, application of conceptual knowledge, communication skills, awareness of limits of knowledge and autonomy and professional capacity" (Connolly, Taylor, & Stone, 2007, p. 2). These external reviewers also remarked on the competencies and capabilities in formulating researchable questions, information literacy, critical thinking, and evidence-based and persuasive communication as invaluable and transferable to other contexts that would serve the student well.

The reviewers observed a tangible community supporting inquiry. Clear and transparent openness to discussion and dialogue is both encouraged and modeled by support staff and faculty. There is open and frequent communication among instructors of any given inquiry course; but collegial discussion about pedagogy and collaboration is also the norm across the program. It may not matter that only some students notice that instructors practice outside of the classroom the skills that they are expecting in the classroom (Connolly et al., 2007, p. 1), because these aptitudes are foundational characteristics of well-functioning organizations. Students certainly do notice that they evaluate their instructors formatively at mid-term as well as summatively at the end of the course, although they may not know they are using the criteria of effective undergraduate instruction identified by Chickering and Gamson (1987). Consequently, while there may occasionally be miscommunications between students and instructors, the attention to program and course structure reported by the review team (Connolly et al., 2007, p. 1) ensures that there are no hidden expectations or "bait and switch" misalignments of desired outcomes and instructional activities. As unfamiliar and ambiguous as inquiry is to the uninitiated, students quickly learn that they actually will be supported in and held accountable for progressive, but genuine, self-direction.

The 2007 program review observed a high degree of congruence between the complex adaptive system students were being prepared for (their experience of the post-graduation real world) and the complex adaptive system that is the BHSc program (Connolly et al., 2007, p. 9). Students authentically hold the reins of their learning, within the context of the content objectives. Each student cohort is contributing to and co-creating the program in which they study.

Although the success of students over the years is a powerful indicator of the performance of the program,[6] there are a number of within-course results to note. The key resources are Trim's (2006) dissertation and Barrett et al.'s (2005) evidence for the prestigious Alan Blizzard Award, given to the Level I Inquiry instructor team for exemplary collaboration to enhance student learning.[7] The key program outcome that Trim found was superior student confidence in learning: BHSc students were far more aware of their ability to locate, access and evaluate appropriate resources, and to evaluate and communicate academic information (Trim, 2006, p. 171). Significant use of small group inquiry is justified because Trim found that her non-inquiry control group of first-year students showed, over time, a reduced preference for learning for understanding (Trim, 2006, p. 147) and a lower ability to appraise arguments, think critically, self-assess their progress and stay organized in their learning (Trim, 2006, p. iv). These findings appear somewhat worse than Springer et al.'s (1999) finding, in which a non-inquiry instructional approach showed comparatively lower proficiency in accurate renditions of the concepts, without lowered gains in knowledge, relative to their small group inquiry treatment.

The process of asking good inquiry questions develops that proficiency and open new pathways for thinking about problems. Although it is apparently a vital part of inquiry (Faculty of Health Sciences, 2014; Justice et al., 2001), questioning was not a skill that Trim (2006, p. 170) considered to be fully formed by the end of a first course in inquiry. This is perhaps not surprising when taken in combination with her observation that these are complex skills (Trim, 2006, p. 167). Asking good researchable questions may be the pinnacle of intellectual expertise, seen at its best in exceptional professional researchers. The subtext of Berger's (2014) popular nonfiction book is that proficiency in formulating good questions is valuable but quite rare.

Although developing a scholarly attitude is definitely among the learning goals of inquiry (Barrett et al., 2005, pp. 7–8; Trim, 2006, p. 174), it must be said that some students crave less ambiguity and more structure (Trim, 2006, p. 167). A minority of inquiry students are lost to the inquiry experience, perhaps because they lack confidence in their critical thinking skills or because they maintain a social construction of education centered on a "transmission and memorization" model of learning (Trim, 2006, p. 168, citing Cuneo & Harnish [2002]). Self-direction is decidedly risky and difficult work, involving trusting the process and suspending self-criticism until the ambiguity begins to clear and proficiency asserts itself. Inquiry does not work for all students (Roy, 2007, p. 41), but in my experience teaching using both approaches, far fewer students are lost to inquiry than are lost to uneven understanding from a lecture course.

Student attrition in the BHSc program is the lowest of any program on campus,[8] although typically one in five students applies to graduate with a three-year general degree because they were accepted into medical school before they finished the full four years. Graduates are encouraged to keep in touch – in fact, they are never disenrolled from the classroom management software – and not surprisingly many spontaneously write to share the interesting challenges they have set for themselves. Notwithstanding the culture of connection actively cultivated within the program, these are surely a self-selected group. Trim (2006) addresses issues of selection bias through extensive and diverse data collection to try to triangulate as wide a picture of the inquiry program contrasts as possible.

Education research must continuously confront questions of validity and replicability owing to the reliance on human subject recollection of experiences. However, a second aspect of methodology may need addressing more systematically: we do not know how much of the success of the program is due to the caliber of the students admitted into the program, as distinct from the program itself. Harnish (2012, p. 108) raises this point, also suggesting post-graduate follow-up studies to complement what is already known about the value added of the program itself. Of course, such research would still need to address selectivity. However, if the question is, "does the program support the program's goals effectively?" then the relevant subject pool surely should be the program's students without reference to an external comparator group.

It is certainly time to revisit the qualitative and quantitative assessment of the program. Anecdotally, the outcomes have not changed since the work of Barrett et al. (2005) and students are achieving strong outcomes in their program learning. However, designing the program from an evidence basis and monitoring the outcomes carefully in the earliest years was a prudent inoculation against the vicissitudes of austerity that rationalized the reduction of inquiry in other faculties (although finding enough skilled instructors may have been the true constraint; see Maurer, 2007) and against the diminution of problem-based learning in some primary health sciences professional programs at McMaster. As the program transitions to a well-proven continuing experiment, up-to-date evidence about the main drivers of the key outcomes will be crucial tools for the next generation of innovators.

Lessons Learned in the Program's Evolution and Renewal

A useful list of insights about teaching inquiry is found in Roy (2007). Here, the focus is on lessons for setting up and nurturing an effective undergraduate program organized around integrated collaborative small-group inquiry. The first message is that local budgetary control is valuable because it provides considerable latitude to be responsive to evidence and student feedback. Logically, a program should be judged on its merits, but the academy is a conservative institution and can be resistant to change. Alternative ideas about what instruction should look like can place a program in jeopardy with just one university president or provost change, unless the program director has negotiated the authority to be truly autonomous, with no political strings attached.

A propos of autonomy, the value of collecting and disseminating the evidence on the effectiveness of the program cannot be understated. New and innovative programs take time to mature, and should be permitted to evolve; but the academy is under great pressure to perform, typically to specific outcomes measures (Altbach, Reisberg, & Rumbley, 2009; Council of Ontario Universities, 2011). Researchers have been questioning the effectiveness of traditional instructional approaches for a number of years (Bell, 1966; Bligh, 2000; Kohn, 1986; Wiggins, 1993). Ironically, instructors considering any instructional alternative must convince skeptics that the new approach is valuable, while an instructor rarely needs to defend their use of the familiar transmission-based pedagogic approaches. Anticipating and even instigating that discussion can be useful, particularly if program directors plan to bring a wide range of evidences to the table.

There is considerable merit in a faculty and staff that predicate their pedagogic actions on evidence and collaboration. It can be helpful to retain the freedom to choose people who are skilled at supporting students seeking their own answers. The best inquiry instructors are open to a democratic and supportive learning environment where student input helps shape the program itself. At the same time, it is challenging, and also risky, to remain consistently "beside" rather than "in front" of the students, creating supportive discomfort by advising and constructively questioning rather than giving in to the pressure to take the reins of the course. Because the instructor is likely to grow dramatically from their involvement with the program (Maurer, 2007; Roy, 2007, p. 41), it is important that they be involved by choice. Providing the instructors with fora for development, through retreats and collegial instructor team meetings, has had dramatic value to goal achievement and the congruence of message.

Course instructor evaluations ask students to describe how effectively the instructor demonstrated Chickering and Gamson's (1987) criteria of good practice at least twice in the course. Asking questions about only those aspects of the course that the instructor can change communicates to students the seriousness with which these evaluations are taken. The habit of debriefing formative feedback in class with the students reinforces that credibility; nonetheless, although response rates are consistently significantly higher in my BHSc program courses than they are in my economics program courses, in neither setting do I believe that student evaluations could be argued to be free of selectivity bias.[9] Many across the academy are exploring the appropriate use of student evaluations of teaching to open up a conversation about undergraduate education, but getting a meaningful sample is also important.

CONCLUSIONS

The Bachelor of Health Sciences Honours program has contributed much that is valuable, first by graduating almost 2000 resourceful, adaptable people into a range of postgraduate vocations, and second by providing a useful model for others of an ongoing experiment in evidence-based instruction. By devoting considerable resources to the first year of the student's program time, and providing myriad opportunities to practice new intellectual habits of mind, students integrate the skills of resilience and self-direction, so that in the upper years these capable emerging researchers can apply their diverse critical thinking skills to a range of challenges with minimal instruction, more like authentic research.

An inquiry-based curriculum requires a certain amount of faith in the learning process, as much for the instructor as for the student. Deliberate work is needed to create a learning environment as free as possible from hegemony, where it is possible to cultivate a joyful curiosity without arbitrary judgment. Learning to question in ways that inspire, rather than expose mistakes, can reorient the work of the academy and leverage the resources of the bright young people who trust their education to it.

NOTES

1. This paragraph draws on personal communication with Dr. Harnish; see also Harnish (2012).
2. The evidence of this claim is the extensive reliance on faculty across McMaster University to provide undergraduate research supervision to students in the upper years of the program; and confirmed anecdotally by the author's conversations over the years with faculty outside of the program.
3. Del Harnish, personal communication.
4. Each instructor has a slightly different way of contextualizing the inquiry skills, based on their own experience of inquiry. In talking to instructors, it is clear that the skills themselves are well-understood to be those presented below in Table 1. Roy (2007, pp. 38–39) identifies four skills as central to inquiry: more responsibility for determining what the student needs to learn; more responsibility

for identifying resources and using them to learn; skills in reporting what has been learned (orally and in writing); and the ability to self-assess the above skills.

5. Each fall, no more than one class of new students is accepted at the second year level as transfers who have already achieved some post-secondary credits. These students work together in a Level II version of Level I inquiry, so that they have the requisite facility in foundational inquiry skills by the end of their initial year in the program.

6. "Approximately 70 percent of students enter medical school, and the rest enter a variety of postgraduate programs according to their extrinsic goals" (Harnish, 2012, p. 108). The 2013-2014 graduate survey reports the medical school application rate at just under 50%. See http://fhs.mcmaster.ca/bhsc/documents/gradchart1314.pdf

7. The BHSc first year inquiry instructor team of Barrett & al. (2005) were the second group of inquiry instructors at McMaster University to win this prestigious award. The first team was Justice & al. (2001).

8. Teresa Basilio, Bachelor of Health Sciences program manager, personal communication.

9. Nilson (2012) of course raises a number of more fundamental problems with the validity of student evaluations of teaching, including the lack of relationship between student evaluations of teaching and student learning.

REFERENCES

Altbach, P., Reisberg, L., & Rumbley, L. (2009). *Trends in global higher education: Tracking an academic revolution*. A report prepared for the UNESCO 2009 World Conference on Higher Education. Paris: UNESCO.

Baird, J. R. (1988). Quality: What should make higher education 'higher'? *Higher Education Research and Development*, *7*(2), 141–152.

Barrett, S., Butler, J., Cates, E., deLottinville, C., Harnish, D., Jordana, M., Kustra, E., Landicho, J., Lee, A., McKinnell, J., McLellan, A., Nastos, S., Nifakis, D., Park, S., Ritz, S., Secord, M., Szechtman, H., & Trim, K., (2005). Skill development with students and explicit integration across four years of the curriculum. In *The Alan Blizzard award: An award for collaborative projects that improve student learning*. Toronto: McGraw-Hill Ryerson. Retrieved from http://www.stlhe.ca/wp-content/uploads/2011/06/2005-Blizzard-McMaster.pdf

Bell, D. (1966). *The reforming of general education: The Columbia college experience in its national setting*. New York, NY: Columbia University Press.

Berger, W. (2014). *A more beautiful question: The power of inquiry to spark breakthrough ideas*. New York, NY: Bloomsbury.

Biggs, J. (1996). Enhancing teaching through constructive alignment. *Higher Education*, *32*(3), 347–364.

Bligh, D. (2000). *What's the use of lectures?* San Francisco, CA: Jossey Bass Publishers.

Bloemhof, B. (2007). Teaching social sciences inquiry: It's not about the content! In C. Knapper (Ed.), *Experiences with inquiry learning* (pp. 31–36). Hamilton: Centre for Leadership in Learning, McMaster University.

Brookfield, S. D., & Preskill, S. (1999). *Discussion as a way of teaching: Tools and techniques for democratic classrooms*. San Francisco, CA: Jossey-Bass Publishers.

Candy, P. C. (1991). *Self-direction for lifelong learning: A comprehensive guide to theory and practice*. San Francisco, CA: Jossey-Bass Publishers.

Chickering, A. W., & Gamson, Z. F. (1987). Seven principles for good practice in undergraduate education. *American Association for Higher Education Bulletin*, *39*(7), 3–7.

Connolly, M., Taylor, L., & Stone, J. (2007). *Undergraduate Program Review Report: Bachelor of Health Sciences Program, McMaster University*. BHSc (Honours) Program, Hamilton, Canada.

Council of Ontario Universities. (2011). *Ensuring the value of university degrees in Ontario: A guide to learning outcomes, degree level expectations and the quality assurance process in Ontario*. Retrieved from http://cou.on.ca/reports/ensuring-the-value-of-university-degrees/

Faculty of Health Sciences. (2014). *What is inquiry?* Bachelor of health Sciences program. Retrieved from http://fhs.mcmaster.ca/bhsc/inquiry_what.html

Harnish, D. G. (2012). Learning as a contact sport: Working with larger system issues to experiment with better learning. In J. K. Dorsey & P. K. Rangachari (Eds.), *Students matter: The rewards of university teaching* (pp. 99–111). Springfield, IL: Southern Illinois University School of Medicine.

Hudspith, B., & Jenkins, H. (2001). *Teaching the art of inquiry*. Halifax: Society for Teaching and Learning in Higher Education.

Justice, C., Rice, J., Warry, W., Inglis, S., Miller, S., & Sammon, S. (2007). Inquiry in higher education: Reflections and directions on course design and teaching methods. *Innovative Higher Education*, *31*(4), 201–214.

Justice, C., Warry, W., Cuneo, C., Inglis, S., Miller, S., Rice, J., & Sammon, S. (2001). *A grammar for inquiry: linking goals and methods in a collaboratively taught social sciences inquiry course*. The Alan Blizzard award paper. Retrieved from http://www.stlhe.ca/wp-content/uploads/2011/06/2001-Blizzard-McMaster.pdf

Kenny, R. W. (1998). *Reinventing undergraduate education: A blueprint for America's research universities*.The Boyer Commission on Educating Undergraduates in the Research University. Stoney Brook, NY: State University of New York at Stony Brook. Retrieved from http://www.niu.edu/engagedlearning/research/pdfs/Boyer_Report.pdf

Knowles, M. (1975). *Self-directed learning; A guide for learners and teachers*. Englewood Cliffs, NJ: Cambridge Adult Education.

Kohn, A. (1986). *No contest: The case against competition*. Boston, MA: Houghton Mifflin.

Lee, V. S. (2007). The roots of inquiry learning inn psychology and philosophy. In C. Knapper (Ed.), *Experiences with inquiry learning* (pp. 13–20). Hamilton: Centre for Leadership in Learning, McMaster University.

Maurer, D. (2007). Teaching inquiry at McMaster: The impact on the instructor. In C. Knapper (Ed.), *Experiences with inquiry learning* (pp. 81–88). Hamilton: Centre for Leadership in Learning, McMaster University.

Nilson, L. (2012). Time to raise questions about student ratings. In J. Groccia & L. Cruz (Eds.), *To improve the academy: Resources for faculty, instructional, and organizational development* (Vol. 31, pp. 213–228). San Francisco, CA: Jossey-Bass Publishers.

Postman, N., & Weingartner, C. (1969). *Teaching as a subversive activity*. New York, NY: Dell Publishing Co. Inc., by arrangement with Delacorte Press.

Reynolds, R., Saxon, D., & Benmore, G. (2006). Impact on the student experience of extending problem-based and enquiry-based learning. *Industry & Higher Education*, *20*(5), 359–370.

Roy, D. (2007). Weaving inquiry into the first year experience: A rope of sand. In C. Knapper (Ed.), *Experiences with inquiry learning* (pp. 37–45). Hamilton: Centre for Leadership in Learning, McMaster University.

Slavin, R. E. (1996). Research on cooperative learning and achievement: What we know, what we need to know. *Contemporary Educational Psychology*, *21*(1), 43–69.

Springer, L., Stanne, M. E., & Donovan, S. S. (1999). Effects of small-group learning on undergraduates in science, mathematics, engineering, and technology: A meta-analysis. *Review of Educational Research*, *69*(1), 21–51.

Tough, A. (1971). *The adult's learning projects: A fresh approach to theory and practice in adult learning*. Toronto: Ontario Institute for Studies in Education.

Trim, K., Harnish, D., Lovric, M., & Roy, D. (2007). Adjusting to learning and teaching with inquiry: A qualitative study of McMaster first-year science students and their instructors. In C. Knapper (Ed.), *Experiences with inquiry learning* (pp. 59–71). Hamilton: Centre for Leadership in Learning, McMaster University.

Trim, K. S. (2006). *Developing inquiry skills with undergraduate students at McMaster 1999–2004: Ongoing formative and summative evaluation for the faculties of science and health science*. Unpublished doctoral dissertation, Clinical Epidemiology and Biostatistics, McMaster University.

Wiggins, G. P. (1993). *Assessing student performance: Exploring the purpose and limits of testing*. San Francisco, CA: Jossey-Bass Publishers.

TWO DECADES OF INQUIRY-ORIENTED LEARNING IN FIRST YEAR UNDERGRADUATE PHYSICS LABORATORIES: AN AUSTRALIAN EXPERIENCE

Les Kirkup

ABSTRACT

A review of the first year physics laboratory program in 1991 at the University of Technology, Sydney (UTS) revealed that student laboratory experiences did not: resemble the practice of physicists; give a realistic picture of the contribution of physics to everyday life, or; enhance students' capabilities of broad value, such as their communication skills. Physics academics at UTS committed themselves to reforming students' laboratory experiences with inquiry-oriented learning as a center-piece of the reform. This chapter explores the drivers that led to the reconceptualization of the role of the laboratory in the undergraduate curriculum and the strategies and processes we adopted over more than 20 years to embed inquiry-oriented activities into first year physics laboratory programs.

Inquiry-Based Learning for Science, Technology, Engineering, and Math (STEM) Programs: A Conceptual and Practical Resource for Educators
Innovations in Higher Education Teaching and Learning, Volume 4, 41–58

ISSN: 2055-3641/doi:10.1108/S2055-364120150000004003

INTRODUCTION

There is a renewed emphasis on enhancing student learning in science through having students engage in the processes of inquiry (Etkina et al., 2010). This emphasis is evident through the stress placed on science inquiry skills in years K-12 of the Australian Science Curriculum (National Curriculum Board, 2010) and at tertiary level through the central place of inquiry and problem solving in the Australian Science Learning and Teaching Academic Standards Statement (Jones, Yates, & Kelder, 2011).

It has been recognized for decades that the processes of inquiry should feature prominently in science education (Schwab, 1960). In criticizing experiences routinely offered to students in an undergraduate physics laboratory, Bless (1933, p. 89) remarked:

> Cookbook instructions certainly do not stimulate the student's capacity for reasoning or ingenuity. If anything, these are stifled under such a procedure .... it is time to realise that the results [of experiments] are of much lesser importance than the process by means of which these results are obtained.

Experiments which rely on "cookbook instructions" are highly guided, verification-type experiences or activities (Hegarty, 1978). Students follow detailed instructions in order to reach a well-defined, and often already well-established, end-point. Such experiments were the norm in first year physics subjects at UTS up to 1991. At that time academics in the Department of Applied Physics (DAP) at UTS began to realize opportunities were being wasted to enhance student engagement and learning through actively involving students in the processes of inquiry.

The importance of experimentation and the processes of inquiry in the science curriculum are reflected in documents of national status. For example, the American Association of Physics Teachers (American Association of Physics Teachers [AAPT], 1998) placed "the art of experimentation" at the top of their list of the goals of undergraduate physics laboratories, expressed as: "The Art of Experimentation: The introductory laboratory should engage each student in significant experiences with experimental processes, including some experience designing investigations" (AAPT, 1998, p. 483).

The AAPT elaborated on this goal: "Laboratory investigations should engage students in the process of formulating and asking an interesting question about nature. Students then select the methods and apparatus needed to make progress towards finding an answer" (AAPT, 1998, p. 484).

This chapter describes factors that stimulated a major reconsideration of the first year physics laboratory programs at UTS and the approaches we adopted to revitalize those programs by embedding inquiry into experiments for three large enrollment first year physics subjects:

- Physics 1C, taken by ≈150 students/year majoring in chemistry, physics, or materials science;
- Physical Modelling, taken by ≈600 students/year majoring in engineering, and;
- Physical Aspects of Nature, taken by ≈650 students/year majoring in the medical, biological or environmental sciences.

The reconceptualization of the laboratory program for Physics 1C acted as a touchstone for later developments of Physical Modelling and Physical Aspects of Nature. Accordingly, most attention is paid in this chapter to the redevelopment of Physics 1C.

REVIEW AND DEVELOPMENT OF PHYSICS 1C

In the early 1990s around 150 students enrolled in Physics 1C in their first semester at UTS. The delivery of Physics 1C consisted of, on average, 3.5 hours of lectures/tutorials each week for 13 weeks. Students, working in pairs, completed one experiment of 2.5 hours duration each week for 10 weeks. A laboratory group consisted of up to 18 students supported by a "demonstrator" (equivalent to a teaching assistant in US universities). Each week, some students would be assigned a mechanics experiment, others an electricity experiment, and so on until over a semester they had completed their allocated experiments. They would typically carry out an experiment either well in advance or well in arrears of consideration of the relevant underlying concepts in lectures or tutorials. This represented a systemic disconnect between lectures, tutorials and laboratories that made up each first year subject.

By 1991 several drivers encouraged the DAP to review the Physics 1C laboratory program. These drivers included:

- dissatisfaction with the 1991 laboratory program as expressed by students through anonymous surveys;
- faculty discontent with a laboratory program virtually unchanged for more than a decade;

- a determination to critically review the purpose and delivery of first year laboratory programs prompted by consultations with teaching and learning specialists at UTS; and
- the infectious enthusiasm of a head of department committed to reforming the laboratory experience of first year students.

At the heart of the reform of the Physics 1C laboratories was the simple question: what do we want students to get from laboratory work? To address this question we assembled a 12 member working party (WP) consisting of physics academics, demonstrators, technical staff, educational researcher, and an educational developer.

As we explored this question, it emerged that many facets of laboratory work we valued as practicing scientists, such as opportunities to discuss findings with peers, were absent from the laboratory program. The WP members reviewed the experiments and made judgements as to whether a student, as a result of carrying out experiments in the program, would have to (Gosper, Cheary, Hazel, & Kirkup, 1993):

(a) engage in scientific inquiry by exploring a phenomenon, developing a procedure, or designing an apparatus;
(b) demonstrate experimental skills or use an instrument or device;
(c) use data reduction skills such as graphical or error analysis; and
(d) demonstrate enhanced physics knowledge by doing the experiment.

Of the 14 experiments reviewed, 100% satisfied (b) and (c) above. Only 21% of the experiments required exploration. None required a student to design an apparatus or develop a procedure, and only 7% required students to demonstrate enhanced physics knowledge.

The next stage of the review examined the openness to scientific inquiry of each experiment using the scheme illustrated in Table 1, adapted from Boud, Dunn, and Hegarty-Hazel (1989). The table relates the level of scientific inquiry inherent within an activity to the extent of student autonomy.

An experiment in which students follow instructions to reach an outcome or answer that is already known, is classified a level 0 experiment. At the other extreme, a level 3 experiment is one in which the students devise the aim, the means by which the aim will be achieved, and are responsible for advancing and defending their answer. Patently, student autonomy and responsibility increase as the level rises from 0 to 3. Of the experiments reviewed, 14% (2 experiments) were classified at level 0 and 86% (12 experiments) at level 1. No experiments were classified by the WP at level 2A or higher.

Table 1. Level of Openness in Scientific Inquiry.

Level	Aim (Is the aim of the investigation given to students or is it open?)	Materials (Do students source their own materials or are the materials provided?)	Method (Is the method given or do students develop their own method?)	Answer (Is the outcome known by, or given to, students, or is the outcome open?)
0	Given	Given	Given	Given
1	Given	Given	Given	Open
2A	Given	Given whole or part	Open or part given	Open
2B	Given	Open	Open	Open
3	Open	Open	Open	Open

Feedback from student questionnaires about the Physics 1C laboratory in 1991 indicated that students were preoccupied with the technical requirements of an experiment with little time for, or consideration of, design, analysis or inquiry. These findings, along with an extensive review of the laboratory program by the WP, led to the conclusion that the 1991 laboratory program (Gosper et al., 1993, p. 9):

> was unlikely to enhance students' interest or sense of excitement about physics, unlikely to help motivate students to continue physics studies at a higher level, unlikely to help develop positive attitudes to science or build scientific literacy. Moreover, the laboratories offered an inaccurate picture of physics and its relevance to everyday life.

The WP committed to reform the Physics 1C laboratory program with the aim of making it more stimulating, personally involving and relevant to student lives. Educational research (Welch, Klopfer, Aikenhead, & Robinson, 1981) offered evidence that an inquiry-oriented approach could accommodate those aims as well as expose students to experiences such as *curiosity, perseverance, experiencing failure and dealing with doubts* (Tamir, 1983). The development of the new program drew on the constructivism theory of learning. According to this theory, students create meaning through investigation and engagement often within real world contexts (Lee, 2011). Beyond this, the WP agreed that a new program should offer students opportunities to engage in activities that closely imitate those of practicing scientists. The program recognized that students use prior knowledge to make sense of new information and situations and that meaningful contexts and relevance are keys to student engagement (Ramsden, 2003).

All the experiments within the 1991 Physics 1C laboratory program were set aside and new experiments devised. Six experiments were developed with inquiry as a central theme (Cheary, Gosper, Hazel, & Kirkup, 1995; Gosper et al., 1993). Classification of the experiments using Table 1 indicated that one experiment was at level 1, with the others at level 2A. The new experiments were trialed with students in 1992 and evaluated through staff and student interviews and surveys.

In the revised laboratory program, all students carried out the same experiment in the same laboratory session allowing the experiment to be sequenced with the lectures. It was observed that students working on the same experiment at the same time stimulated meaningful discussions and comparisons of the various approaches adopted.

Student Assessment

The WP recognized that the emphasis brought to inquiry, experimentation, and communication must be reflected in the assessment scheme for the laboratory program. Prior to the development of the post 1991 laboratory program the assessment items were: (i) a formal report of one of the experiments carried out by the student during the semester, and; (ii) an end of semester practical exam in which students performed an experiment then wrote it up during the exam. A diverse assessment scheme was implemented for the Physics 1C laboratory program, consisting of:

- Compulsory prework to be completed ahead of the laboratory. The prework might involve a take-home kit which students use to carry out simple experiments in order to become familiar with some concepts and experimental challenges that feature in the experiment it supports;
- A skills test in which students work individually, requiring students critique a section of a report, perform data analysis and carry out a hands-on experiment (categorized as level 1 using Table 1);
- Examination of a student's laboratory notebook in which a student makes a contemporary record of their ideas, experimental design, data, calculations, difficulties, findings and tentative conclusion(s). This replicates a routine, but core, activity of professional scientists;
- A formal report requiring students to communicate their aim, methods and findings in a manner consistent with discipline norms; and
- A poster and poster presentation. Students cooperate to present an experiment and its findings in a visual format used widely in conferences. Students fashion, present and orally defend their poster.

The variety in the student assessment components incorporated into the new laboratory program served to deliver a broader profile of student achievement than the program it replaced. To reflect the increased emphasis on laboratory-based activities, the contribution of the laboratory to the total assessment of Physics 1C grew from 15% to 25%.

Student Reaction to the Revised Physics 1C Laboratory Program

Student reaction to the new approach was obtained through one-on-one interviews. The interviews were designed to establish students' experiences of the new program as well as the extent to which the program matched their abilities, background and anticipated future needs. Feedback from students included:

> I find that (fewer instructions) are better than someone giving us the instructions and saying, do this and do that. If you have a problem you actually work hard at working it out and you feel you have accomplished something at the end. It's better than someone saying calculate this, calculate that. (Gosper et al., 1993, p. 14)

The new program brought an emphasis to formal and informal written and oral reporting within a supportive, small group laboratory. This was generally viewed positively by students: "The discussion helps to answer the questions that never get asked or answer the questions you wanted to ask and felt too stupid to ask .... Everyone benefits" (Gosper et al., 1993, p. 15).

Experiments in the post 1991 laboratory program emphasized the place of physics in the real world, in contrast to the largely context-free experiments which they replaced. A student repeating Physics 1C offered a comparison between the old and new programs:

> It's giving a much better perspective about what physics is all about. When I started term this year I wasn't sure whether I was interested – but now that the course has turned out the way it has, I'm really enjoying it. (Gosper et al., 1993, p. 16)

The revised Physics 1C laboratory program gave students more control over their experiences in the laboratory by allowing them opportunities to engage in inquiry. In doing so students participated in activities habitually carried out by practicing scientists which has been shown to promote positive attitudes toward science and enhance comprehension (Weaver, Russell, & Wink, 2008).

The revitalization of the laboratory program for Physics 1C students encouraged the DAP to revise the laboratory program for UTS engineering students required to enroll in the first year physics subject "Physical Modelling."

REVIEW AND DEVELOPMENT OF PHYSICAL MODELLING

Our plan to reform our physics subject for engineering students called Physical Modelling occurred subsequent to a review of engineering education carried out by the engineering profession in Australia. An outcome of the review was a report entitled *Changing the culture: engineering education into the future* (Institution of Engineers, Australia, 1996) which articulated the desirable attributes of engineering graduates, including the ability to:

- apply knowledge of basic science and engineering fundamentals;
- communicate effectively, not only with engineers, but also with the community at large;
- undertake problem identification, formulation and solution.

It was against this backdrop that we revitalised the Physical Modelling laboratory program (Kirkup et al., 1998). Supported by an internal UTS teaching and learning grant, a team consisting of physics and engineering academics, an educational developer, and several technical staff was assembled. The aim of the development was to bring a balance to laboratory work focussing on specific technical competences and more open activities relevant to engineers which foster independent learning, critical thinking and professional values.

Throughout the process of reform of the Physical Modelling laboratory program we canvassed a wide range of stakeholders. We identified, consulted and in some cases surveyed: first year engineering students, lecturers from the Faculty of Engineering at UTS, Engineering policy makers at UTS (including the Dean of Engineering, Associate Deans and Heads of School), senior committee members from the peak professional engineering body in Australia, namely the Institution of Engineers, Australia.

One-on-one interviews took place with academics and engineering professionals. The interviews, which were audio-recorded, were typically of 30 minutes. They were then transcribed. Some of the interviews were face to face while others were carried out by telephone. We asked:

- Ideally, what qualities should undergraduate [physics] laboratory programs develop in engineering students?
- Which of these [qualities] could begin to be developed in a first year [physics] laboratory program?

An emphatic response that emerged from the consultations was that the laboratory program should put physics first and applications to engineering second (Kirkup et al., 1998). On the more specific matters regarding physics laboratory program and the qualities such a program might develop in students (Kirkup et al., 1998, p. 260):

> [physics labs should develop] ability in problem formulation: thinking about equipment which might be used to address or explore the problem. (Dean of Engineering Advisory Committee member, UTS)

Student views on the existing laboratory program were sought through a questionnaire (Kirkup et al., 1998). Students believed the program could be enhanced by coordinating the lecture and the laboratory programs and that the experiments should be made more relevant to engineering and/or everyday life. The importance of sequencing lectures and laboratories was also recognized and commented upon by the engineering academics surveyed. As one senior engineering academic remarked (Kirkup et al., 1998, p. 260): "Surely in the early years lectures and labs should be sequenced – what do we think the students make of it otherwise?"

Consultations with stakeholders, and consideration of the teaching and learning literature allowed us to frame a set of aims against which the revised laboratory program could be evaluated. We determined that the laboratory program should have relevance to students' professional development by: having an engineering perspective; experiments that relate to real world problem(s) or issue(s); encouraging the development of experimental design skills; needing careful measurements; utilizing data analysis techniques; involving the comparison of models with experimental data, and; requiring students to engage in effective teamwork.

As with Physics 1C, we determined to infuse a strong inquiry-oriented focus into the Physical Modelling laboratory program. This led us to set aside the experiments that engineering students were required to perform up to 1996. Up to 1996 the experiments were essentially recipe-based with no engineering context. Students worked in pairs with little opportunity to develop their own methods or discuss their results with other students. Post 1996, new experiments were created with an overt engineering slant that would be classified by Table 1 as level 2A or 2B.

The redesigned laboratory program included:

- experiments with an engineering flavor;
- compulsory pre-work accompanying each laboratory session;
- time for students to devise and carry out their own experimental procedures;
- opportunities to describe their methods and results to the whole class in a semi-formal manner;
- experiments linked to and sequenced with, material delivered in lectures;
- larger units of work spanning more than one week, and;
- a diverse assessment scheme.

Evaluation of the Physical Modelling Laboratory Program

The laboratory program was evaluated using a student questionnaire and focus groups with engineering students. Details of the result of the questionnaire can be found elsewhere (Kirkup et al., 1998). Findings from completed student questionnaires included:

- Improved sequencing of the lab-based activities made a significant impact on students' awareness of the links between the material in lectures and the experiments;
- The introduction of experiments with engineering context enhanced students' perception of the relevance of the physics experiments to their engineering education;
- Students reported that the laboratory program assisted in improving their understanding of physics concepts and increased their interest in physics; and
- The laboratory program increased students' confidence in doing physics.

The findings of the focus groups, which consisted of between 5 and 10 students chosen at random, largely supported the findings from the student questionnaires. Responses to the question "What did you like most about the laboratory program?":

- "It was really exciting when we did things related to engineering, like the thermoelectric coolers."
- "I found it much better than previous labs, because before teachers would just tell us what to do and we would just do it. But here we would design our own experiments, get the results and use them, and that's much better." (Kirkup et al., 1998, p. 264)

Focus groups revealed that students valued the opportunity to engage in devising a practical solution to a problem that had meaning for them while being able to connect theory with practice. The readiness of students to grasp the opportunities that inquiry-oriented activities have to offer was highly variable between students. This is a situation familiar to those introducing inquiry in the laboratory at all levels of education (Wenning, 2005).

Successful revision of the physics laboratory programs for engineering students encouraged us to turn our attention to the physics subject called Physical Aspects of Nature which enrolled students intending to major in the medical, environment and biological (MEB) sciences at UTS.

DEVELOPMENT OF PHYSICAL ASPECTS OF NATURE (PAN)

A major redevelopment of the laboratory program for PAN which occurred post-2000 allowed us to integrate and apply what was learned developing laboratory programs for physical science and engineering majors. As with the development of the laboratory program for Physics 1C and Physical Modelling, the emphasis remained on students engaging in inquiry in contexts meaningful to them. The development team consisted of experts in pedagogy and other with first-hand experience of working with students in undergraduate laboratories. Views of a wide range of stakeholders were canvassed as the new program was developed and trialed. Successful elements from the development of Physics 1C and Physical Modelling were adopted, adapted and extended.

In 2007 we devised a framework to assist in the more systematic development and evaluation of individual inquiry-oriented experiments (Kirkup, Pizzica, Waite, & Mears, 2012). The framework, which will now be described, is shown in Fig. 1.

In developing new experiments for PAN, we began by identifying topics that could form the basis of inquiry-oriented experiments. This occurred through attendance of a member of the laboratory development team at lectures given to first year students by faculty from the MEB sciences. This was followed up by one-on-one interviews with the lecturers. Such engagement allowed themes in the curriculum suitable for treatment through inquiry-oriented experiments to emerge that would be difficult to identify solely by inspecting paper-based or online materials provided to students. For example, attendance at lectures indicated that fluid flow was a theme of importance to students from the MEB sciences and so formed the basis of an IOL activity (Kirkup, Pizzica, Waite, & Srinivasan, 2010).

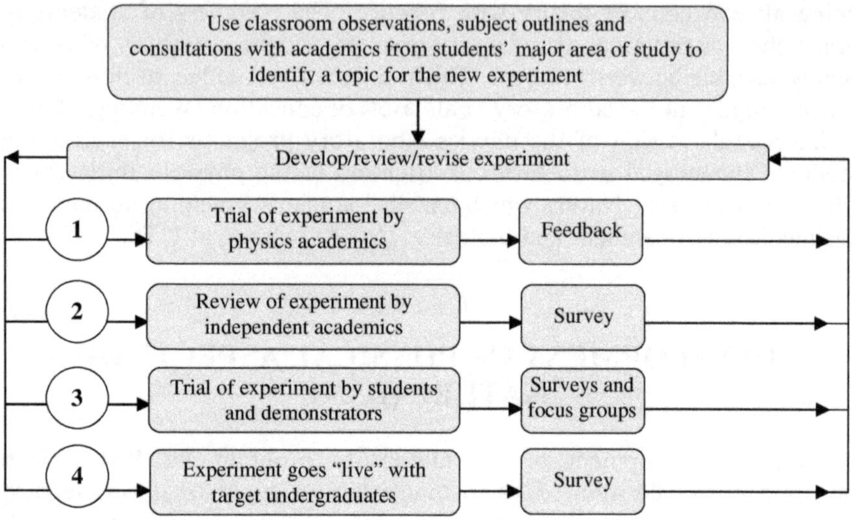

Fig. 1. Framework for Developing and Evaluating an Inquiry-Oriented Laboratory Program.

Steps 1–4, as shown in Fig. 1, inform the development and improvement of the experiment through feedback. More specifically:

- In step 1, an academic trials the experiment considering its feasibility (for example with respect to time required to complete the experiment), the equipment requirements, and how the students should be prepared for the experiment (for example through carrying out a pre-laboratory exercise).
- In step 2 the written materials created to assist students in the laboratory are reviewed, not only by physics academics who could assure that the experiment is neither too demanding nor too trivial, but also by faculty from the MEB sciences. This allows faculty to advise on the relevance of the experiment to students majoring in their disciplines.
- In step 3, the experiments are trialed with a group consisting of students who have already completed PAN in an earlier semester and physics demonstrators not involved in the design of the new program. Students are paired with demonstrators to work on the experiment under the guidance of a senior demonstrator who was one of the developers of the experiment. Students and demonstrators are surveyed on the completion of the experiment and also take part in focus groups, and;

- In step 4, the experiments are rolled out to the target audience and feedback gathered through surveys and focus groups.

The framework in Fig. 1 promotes an informed and student-centered approach to the development and evaluation of laboratory-based learning activities. It reveals factors likely to affect student engagement allowing for timely intervention by the innovators. The framework promotes the accumulation of evidence about aspects of an experiment, for example whether students regard the experiment as enhancing their capacity to communicate in the laboratory. We learned through the application of the framework that professional development of demonstrators needed further emphasis if opportunities for enhancing student capabilities through inquiry-oriented laboratory work were to be effectively exploited.

DISCUSSION

Drivers for Change Toward Inquiry in the Curriculum

Calls for a greater emphasis on science students engaging in the processes of inquiry have grown over many decades (Tamir, 1983). While the impact on actual practices has been limited, a number of factors are acting to promote such engagement on a large scale in primary, secondary and tertiary education. Those factors include a shared commitment to promoting and embedding inquiry in the curriculum from national and institutional policy makers and those that engage directly with students on a day-to-day basis (Boyer Commission on Educating Undergraduates in the Research University, 1998). More specifically, the drivers for change include:

- Senior figures, including Australia's Chief Scientist are vocal in advocating students be given insights into the processes by which scientific knowledge is created and challenged, by engaging in the process of inquiry themselves (Office of the Chief Scientist, 2013).
- Evidence accumulated over many years that active learning strategies increase student performance and enhance student critical thinking, written and oral communication skills and the capacity to work productively in groups (Hanif, Sneddon, Al-Ahmadi, & Reid, 2009).
- The potential of inquiry skills to ameliorate the decline in STEM participation and arrest the attrition that occurs in STEM courses (Pitkethly & Prosser, 2001).

- Finally, there is increased awareness that inquiry as carried out by undergraduates is able to imitate the research activities of practicing scientists and so is likely to gain favor with academics with a strong research orientation (Healey & Jenkins, 2009).

Making it Happen: Lessons Learned over 20 Years

The pre-1991 laboratory program for first year students enrolling in Physics 1C at UTS contained experiments that included step by step instructions, encouraged very little independent investigation by students, did not link to any everyday applications of physics, and were out of sequence with concepts or principles introduced and discussed in lectures or tutorials. Bringing students, discipline experts and educational researchers together allowed for an informed critique of the Physics 1C laboratory program. This critical mass of participants with varied backgrounds and understandings of physics and science education was vital as it tapped into diverse and discriminating views essential to the development of a successful inquiry-oriented laboratory program. Strong and enthusiastic leadership dedicated to the development was equally important, as it assured the innovation remained a high priority for the DAP until the new program had become established.

Through fortnightly meetings that occurred over a period of approximately 18 months we were able to explore the shortcomings in the current program, and what elements a reconceptualised Physics 1C program should contain. This gestation period allowed a shared vision of the new program to emerge. Without a shared vision of the nature of what is to be achieved, and the means by which it may be achieved, ambitious innovations can easily founder (Hernandez, Ravn, & Forero-Shelton, 2014). The procedure by which the Physics 1C laboratory program was developed, in which of teams consisting of members with a range of backgrounds and interests collaborated with stakeholders was so successful that it was adopted for the development of the Physical Modelling and Physical Aspects of Nature laboratory programs.

In introducing inquiry-oriented activities for all our first year physics subjects, we chose not to modify existing experiments, but set them aside and in the process discard the associated laboratories manuals. Advantages of this were that we could conceptualize the program as a whole rather than a series of isolated episodes students experience while in the laboratory. We were also able to scaffold the learning experiences throughout the semester, beginning with more guided experiments progressing to situations in which students were given more freedom and responsibility to design and carry out their own experiments. As all students worked on the same experiment in the same laboratory period, we were able to sequence the experiments with material introduced in the lectures, bringing a unity and coherence to the subjects.

The innovations introduced since 1991 were prompted by the literature on learning through inquiry (Tamir, 1983) as well as the teams' recognition of the importance of giving students an authentic experience of acting in the role of scientist (Boud et al., 1989). The development process, as embodied in Fig. 1, matured over time as we learned more about involving stakeholder at all stages. Consistent with the findings of others, we found that active learning, in this case learning through hands-on inquiry, increased student satisfaction and engagement (Ahlfeldt, Mehta, & Sellnow, 2005).

Not all academics within the DAP were involved in developing the new laboratory programs but we understood the importance of having the majority of faculty positively disposed toward the revitalization of the laboratory programs. To this end whole of department workshops in which the goals of the new laboratory program were explored and new experiments trialed by full time staff did much to get the majority of staff "on side." Academics not directly involved with the innovations could bring their perspectives of the new program. Perhaps inevitably there were academics (few in number) who could see little value in changing the programs. The input of such academics was vital as a reality check and for us to ask ourselves "are we convinced of the value of the approach(es) we are adopting?"

Supporting Student Learning

In the development of new inquiry-oriented experiments we recognized the importance of setting the experiments within contexts that have meaning for students and to which students bring experiences and attitudes to learning and their beliefs about what constitutes learning. These experiences and attitudes can vary enormously, especially in large enrollment first year subjects that include students from many science disciplines.

It was recognized early that changing the nature of the experiments to include more inquiry required a change in the role and contribution of those supervising the laboratory from instructor to facilitator. Such a change required a shift in the approach to professional development of a strategically important group of individuals. Members of the group going forward would be (on the whole) casual academics, many with study priorities of their own, such as completing a PhD. We introduced demonstrator orientation sessions in which issues that emerge while supporting students learning during an inquiry session could be explored. These sessions were also used to trial new experiments, with the demonstrators as surrogate students. This allowed the demonstrators to become familiar with experiments, examine how they might support student learning when they supervise a class, while at the same time giving feedback as to how the experiments could be improved.

As direct one-on-one interactions between students and demonstrators is largely limited in physics to the laboratory, the extent to which the alignment of the backgrounds, attitudes and experiences of instructors with their students impact on student learning, engagement and satisfaction deserves to be an area of research (Alkaher & Dolan, 2011).

CONCLUSION

Over two decades ago, the DAP at UTS made a determined move to embed inquiry-oriented activities into the laboratory program for students enrolling in a first year physics subject. Since that time inquiry-oriented activities have been expanded to all first year physics subjects, such that to date of the order of 15,000 students have engaged in such activities at UTS. Our experiences of developing inquiry-oriented experiments as described in this chapter lead us to conclude that:

- Inquiry-oriented experiments are superior to recipe-based experiment at providing students with experiences that closely imitate the processes of science and the activities of practicing scientists.
- Context and relevance are vital for student engagement and should be built into inquiry-oriented experiments. This is especially important for the students not intending to major in physics.
- Students expect clears links to be made between their experiments, the wider curriculum and the world outside the laboratory.
- Reforming a laboratory program profits from having a development team with diverse backgrounds and experiences that reaches a common understanding of the purpose of the reform.

Challenges to embedding and sustaining IOL activities have largely been met though the increase in student enrollment means that we must continue to innovate, and most importantly support our demonstrators, as they in turn strive to support student learning through inquiry-oriented activities.

ACKNOWLEDGMENTS

I wish to thank the following people who made major contributions to the design, delivery, evaluation and evolution of the laboratory programs discussed in this paper. David Green, Maree Gosper, Elizabeth Hazel, Sue Johnson, Nirmala Maharaj, Jenny Pizzica, Mark Smith, Lakshmi Srinivasan, Paul Swift, and Katrina Waite. I would particularly like pay tribute to the late Bob Cheary. Without his drive and leadership the first year physics laboratory programs at UTS might still be languishing in the mid-20th Century.

REFERENCES

Ahlfeldt, S., Mehta, S., & Sellnow, T. (2005). Measurement and analysis of student engagement in university classes where varying levels of PBL methods of instruction are in use. *Higher Education Research and Development*, *24*(1), 5–20.

Alkaher, A., & Dolan, W. (2011). Instructors' decisions that integrate inquiry teaching into undergraduate courses: How do I make this fit? *International Journal for the Scholarship of Teaching and Learning*, *5*(2), 1–24.

American Association of Physics Teachers. (1998). Goals of the introductory physics laboratory. *American Journal of Physics*, *66*, 483–485.

Bless, A. A. (1933). Cook-book laboratory work. *American Journal of Physics*, *1*, 88–89.

Boud, D., Dunn, J., & Hegarty-Hazel, E. (1989). *Teaching in laboratories*. Milton Keynes: Open University Press.

Boyer Commission on Educating Undergraduates in the Research University. (1998). *Reinventing undergraduate education: A blueprint for America's research universities*. Stony Brook, NY: State University of New York-Stony Brook. Retrieved from http://www.niu.edu/engagedlearning/research/pdfs/Boyer_Report.pdf

Cheary, R., Gosper, M. V., Hazel, E., & Kirkup, L. (1995). Revitalising the first-year physics laboratories at the University of Technology, Sydney. *Australia & New Zealand Physicist*, *32*, 119–124.

Etkina, E., Karelina, A., Ruibal-Villasenor, M., Rosengrant, D., Jordan, R., & Hmelo-Silver, C. E. (2010). Design and reflection help students develop scientific abilities: Learning in introductory physics laboratories. *Journal of the Learning Sciences*, *19*(1), 54–98.

Gosper, M., Cheary, R., Hazel, E., & Kirkup, L. (1993). Reforming the Physics Laboratory: From Theory to Practice. *The Proceedings of the Third International Seminar on*

Misconceptions and Educational Strategies in Science and Mathematics, Misconceptions Trust: Ithaca, NY. Retrieved from http://www.mlrg.org/proc3pdfs/Gosper_Physics.pdf

Hanif, M., Sneddon, P. H., Al-Ahmadi, F. M., & Reid, N. (2009). The perceptions, views and opinions of university students about physics learning during undergraduate laboratory work. *European Journal of Physics*, *30*, 85–96.

Healey, M., & Jenkins, A. (2009). Developing undergraduate research and inquiry. Retrieved from www.heacademy.ac.uk/assets/documents/resources/publications/%20developing undergraduate_final.pdf

Hegarty, E. H. (1978). Levels of scientific enquiry in university science laboratory classes: Implications for curriculum deliberations. *Research in Science Education*, *8*, 45–57.

Hernandez, C., Ravn, O., & Forero-Shelton, M. (2014). Challenges in a physics course: Introducing student-centred activities for increased learning. *Journal of University Teaching and Learning Practice*, *11*(2). Retrieved from http://ro.uow.edu.au/jutlp/vol11/iss2/8

Institution of Engineers, Australia. (1996). *Changing the culture: Engineering education into the future*. Barton, Australian Capital Territory: Institution of Engineers, Australia.

Jones, S. M., Yates, B. F., & Kelder, J.-A. (2011). *Learning and teaching academic standards project: science learning and teaching academic standards statement*. Sydney: Australian Learning and Teaching Council.

Kirkup, L., Johnson, S., Hazel, E., Cheary, R. W., Green, D. C., Swift, P., & Holliday, W. (1998). Designing a new physics laboratory programme for first year engineering students. *Physics Education*, *33*, 258–265.

Kirkup, L., Pizzica, J., Waite, K., & Srinivasan, L. (2010). Realizing a framework for enhancing the laboratory experiences of non-physics majors: From pilot to large-scale implementation. *European Journal of Physics*, *31*, 1061–1070.

Kirkup, L., Pizzica, J., Waite, K. M., & Mears, A. (2012). *Adaptable resource kit*. Retrieved from http://www.iolinscience.com.au/wp-content/uploads/2011/11/ARK__June-10__2013.pdf

Lee, V. S. (2011). The power of inquiry as a way of learning. *Innovative Higher Education*, *36*, 149–160.

National Curriculum Board. (2010). *Shape of the Australian curriculum: Science*. Retrieved from www.acara.edu.au/verve/_resources/Shape_of_the_Australian_Curriculum.pdf

Office of the Chief Scientist. (2013). *Science, technology, engineering and mathematics in the national interest: A strategic approach*. Canberra: Australian Government: Retrieved from http://www.chiefscientist.gov.au/wp-content/uploads/STEMstrategy290713FINALweb.pdf

Pitkethly, A., & Prosser, M. (2001). The first year experience project: a model for university change. *Higher Education Research and Development*, *20*, 185–198.

Ramsden, P. (2003). *Learning to teach in higher education* (2nd ed.). Oxford: Routledge.

Schwab, J. J. (1960). Inquiry, the science teacher and the educator. *School Review*, *68*, 176–195.

Tamir, P. (1983). Inquiry and the science teacher. *Science Education*, *67*(5), 657–672.

Weaver, G. C., Russell, C. B., & Wink, D. J. (2008). Inquiry-based and research-based pedagogies in undergraduate science. *Nature Chemical Biology*, *4*(10), 577–580.

Welch, W. W., Klopfer, L. E., Aikenhead, G. S., & Robinson, J. T. (1981). The role of inquiry in science education: Analysis and recommendations. *Science Education*, *65*(1), 33–50.

Wenning, C. J. (2005). Minimizing resistance to inquiry-oriented science instruction: The importance of climate setting. *Journal of Physics Teacher Online*, *3*(2), 10–15.

ARNOLD ARONS: PIONEER OF INQUIRY-BASED LEARNING IN PHYSICS☆

Richard R. Hake

ABSTRACT

Arnold Arons, along with Robert Karplus, can fairly be called one of the founding fathers of U.S. Physics Education Research and a pioneer of inquiry methods of education. The instructional methods advocated by Arons were influenced by the work of Socrates, Plato, Montaigne, Rousseau, Dewey, Whitehead, and Piaget, but are primarily derived from Arons' epic half century effort to improve introductory science

☆In this chapter I indicate (a) quotes by indents and smaller font size, (b) my inserts into quotes by "... [[insert]] ..."], (c) URLs by <URL> where the angle brackets serve to promote hot-linking across line breaks and to delineate what *is* and what *is not* part of the URL.

For Arons' obituaries see Haskell (2001), Matthews (2001), McDermott, Wilson, and Jossem (2001), and Woods Hole Oceanographic Institution (2001). For further bibliographic information and a fascinating interview see Minstrell (1981).

"Operational Definitions" are defined below in "Arons-Advocated Method – Emphasis On."

Inquiry-Based Learning for Science, Technology, Engineering, and Math (STEM) Programs: A Conceptual and Practical Resource for Educators
Innovations in Higher Education Teaching and Learning, Volume 4, 59–72

ISSN: 2055-3641/doi:10.1108/S2055-364120150000004005

teaching by shutting up and listening carefully to students' responses to probing Socratic questions on physics, science, and ways of thinking. Arons emphasized: (1) conceptual understanding, (2) operative knowledge, (3) interactive engagement, (4) Socratic dialogue, (5) attention to cognitive development, (6) attention to preconceptions of beginning students, (7) operational definitions, (8) reduction of volume and pace of standard introductory courses, (9) idea first, name afterward, (10) importance of a course "story line," and (11) science as a liberal art. Most of these are attributes of enlightened inquiry-based learning as described in Inquiry and the National Science Education Standards: A Guide for Teaching and Learning (NRC, 2000).

INTRODUCTION

In 1980, after a disastrous attempt to explain Newtonian mechanics to prospective elementary teachers, I sought advice on efficacious physics pedagogy by telephoning around the country. I had little success until science-education pioneer Robert Karplus at Berkeley advised me to call "the only person in the country who understands the problem," Arnold Arons at the University of Washington. In a watershed 30-minute telephone conversation, Arons, speaking from decades of hard-won experience with elementary education majors (Arons, 1977), recommended that I abandon the traditional passive-student lecture. He patiently explained his physics education method: hands-on laboratory experience with concrete physical systems, repeated interactive engagement at increasingly sophisticated levels, emphasis on operational definitions, and Socratic Dialogue.

The Arons-Advocated Method

Seeking to learn more about Arons' unorthodox approach, I studied some of his insightful articles (Arons, 1973, 1974, 1976) and was deeply impressed. I tried to construct several laboratory write ups in an Arons mode, and found that such "Socratic-Dialogue-Inducing" (SDI) labs were relatively effective in advancing the conceptual understanding of physics, not only among elementary education majors (Hake,1991a), but also among pre-medical students and science majors (Hake, 1987, 1991b, 1992, 1998a, 1998b, 2002a, 2002b, 2002c, 2003, 2012), and even non-physical-science professors (Tobias & Hake, 1988). For a more detailed account see "My Conversion to the Arons-Advocated Method of Science Education" (Hake, 1991a).

In the 1980's I attempted to spread the word regarding the efficacy of the "Arons Method" by traveling about the country giving talks with that title. However, I later changed the title to the "Arons-Advocated Method" (AAM) when Arons insisted that he had done nothing original, and that most of what he advocated was from Socrates, Plato, Montaigne, Rousseau, Dewey, Whitehead, and Piaget. Arons was doubtless influenced by those educators, and his method may well be consistent with their work. But, judging from Arons' own words (see below), his method is better described as empirically derived by shutting up and listening carefully to students' responses to probing Socratic questions on physics, science, and ways of thinking. For example, in "Toward Wider Public Understanding of Science: Addendum," Arons (1974) wrote:

> If a teacher disciplines himself to conduct such Socratic Dialogues he begins to see the difficulties, misconceptions, verbal pitfalls, and conceptual problems encountered by the learner … In my own case, *I have acquired in this empirical fashion virtually everything I know about the learning difficulties encountered by students* … [[My *italics*.]] … I have never been bright enough or clairvoyant enough to see the more subtle and significant learning hurdles a priori.

In my opinion, the AAM, with its basis in the astute probing and analysis of student minds by a discerning research physicist deserves greater attention. I would extend Knight's (1995) recommendation that "every teacher of introductory physics, from high school through the calculus-based course, should have a copy of this book … (Arons, 1990, 1997a) … within easy reach," to include every science teacher, education researcher, and cognitive scientist. Judging from (a) psychologist Robert (Grossman's, 2005, 2009a, 2009b) use of Arons' work to elucidate "knowledge transformations," and (b) Tobias and Hake's (1988) AAM for non-science professors, even non-physical science students can benefit.

Arons (1984b) ends his Physics Teacher trilogy (Arons, 1983, 1984a, 1984b) on "Student Patterns of Thinking and Reasoning" with the anticipation that cognitive research will bring a rationale and an order to his abundant empirical insights:

> I fully anticipate that future advances in cognitive research will analyze, reconstruct, merge, separate, and reassemble in various more orderly ways, the insights I have attempted to describe. For the time being I simply present what I have been able to discern through the still surrounding haze.

In the ensuing three decades – cognitive scientists, biologists, neuroscientists, and education researchers have indeed "analyzed, reconstructed, merged, separated, and reassembled" in various ways, ideas compatible with Arons' ethnographic (Howe, 2001; Howe & Eisenhart, 1990) insights and suggestions for the improvement of science education. For example:

(a) cognitive scientists: *How People Learn: Brain, Mind, Experience, and School* (Bransford, Brown, & Cocking, 2000)
(b) biologists: *The Art of Changing the Brain: Enriching the Practice of Teaching by Exploring the Biology of Learning* (Zull, 2002)
(c) neuroscientists: "The Brain. Use It or Lose It" (Diamond, 1996) and
(d) education researchers:
  (i) *RL-PER1: Resource letter on physics education research* (McDermott & Redish, 1999);
  (ii) "Learning as Biological Brain Change" (Leamnson, 2000);
  (iii) "Scientific Research in Education" (Shavelson & Towne, 2002);
  (iv) "Lessons From the Physics Education Reform Effort" (Hake, 2002b);
  (v) "Points of View: On the Implications of Neuroscience Research for Science Teaching and Learning: Are There Any?" (Lawson, 2006); and
  (vi) "Physics Education Research Papers on the Web" (UMPERG, 2015).

What are the salient features of the Arons-Advocated Method? Herewith is my own selection of eleven major components, as culled from my reading of Arons' thoughtful articles and books, listening to his provocative talks, and exchanging opinions on matters of pedagogy and physics. The items included are not: (a) exhaustive – they omit some facets of Arons' strategies, (b) independent – they overlap one another in mutually reinforcing ways, or (c) unique – other physicists and educators might choose to emphasize other components of the Arons method.

THE ARONS-ADVOCATED METHOD

Emphasis on:

1. Conceptual Understanding
2. Operative Knowledge
3. Interactive Engagement
4. Socratic Dialogue
5. Attention to Cognitive Development

6. Attention to Preconceptions of Beginning Students
7. Operational Definitions
8. Reduction of Volume and Pace of Standard Introductory Courses
9. Idea First, Name Afterward
10. Importance of a Course "Story Line"
11. Science as a Liberal Art

In practice:

Arons practiced what he preached. In presenting him with the Oersted Medal for 1972, in honor of his notable achievements in physics teaching, Donnally (1973) wrote:

> While at Amherst College he was widely know as a skilled teacher and was featured in a cover story on education in Time magazine. The content of the course for which he was cited and the instructional philosophy on which it was based are exemplified in his text, *Development of Concepts of Physics*, ... [[Arons, 1965]] ... The very careful attention of a logical sequencing of ideas, the deep concern for a careful development of concepts in the minds of students, and the steady attention to the cultural basis of Western science so evident in that book have been his hallmarks.

Did the Amherst students of the 50s appreciate Arons? In an interview with Arons, his former doctoral student Minstrell (1981) asked him "What has been the highlight of your teaching career?" Arons answered:

> I suppose it was what happened during my first year at Amherst, where I moved from Stevens in the Fall of 1952 ... In the last lecture, instead of a passive review or the addition of more subject matter, I stood back and talked about the meaning of education, about the whole intellectual experience we had tried to make available, about thinking, learning, and the development of their own minds. I pointed out some of the insights they should now have into science as a human endeavor requiring intellect and imagination, a two-edge blade with powers for both good and evil, encumbered with profound limitations. As I ended the lecture, closed my notebook, and was about to leave the room, the class burst into a standing ovation. This was completely unanticipated; I had never had it happen before; it came completely out of the blue. I just gasped, and I can recall a bit of moisture in my eye.

In regard to the liberal side of science/math education, I once asked Arons for his opinion of the virtually forgotten Benezet Method (Benezet, 1935/1936; Mahajan & Hake, 2000) of delaying drill on arithmetic algorithms until the sixth grade in favor of reading; inventing; discussing stories and problems; estimated lengths, heights, and areas; and finding and interpreting numbers relevant to their lives. Arons (2000) responded (my italics):

> I have looked at the Benezet papers, and I find the story congenial. The importance of cultivating the use of English cannot be exaggerated. I have been pointing to this myself since the 50's, and am delighted to find such explicit agreement. I can only point out

that my own advocacy has had no more lasting effect than Benezet's. You will find some of my views of this aspect in ... [["Structure, methods, and objectives of the required freshman calculus-physics course at Amherst College (Arons, 1959)]] ... *Benezet taught excellent arithmetic from the very beginning just as it should be taught*. What he removed was useless drill on memorized algorithms that had no connection to experience and verbal interpretation ... This, of course, brings us back to the same old problem: *Whence do we get the teachers with the background, understanding, and security to implement such instruction*. They will certainly not emerge from the present production mills.

SUMMARY AND CONCLUSIONS

The above eleven aspects of the Arons-Advocated Method are trenchantly summarized in the abstract of "Toward wider public understanding of science" where Arons (1973) wrote:

Wider understanding of science will be achieved only by giving students a chance to synthesize experience and thought into knowledge and understanding. Such a chance is not available in the deluge of unintelligible names and jargon precipitated at unmanageable pace and volume in so large a proportion of our college courses, and it is not available in the absence of humanistic, historical, or philosophical perspectives within those courses. Neither will salvation be found in topical courses on currently "popular" matters such as the energy crisis, environmental problems, or societal impact – so long as these problems are plunged into without any genuine prior understanding of the underlying scientific ideas. Some deeply valuable lessons about the learning process, as it occurs in all individuals, are to be learned from among the best of the elementary science curricula that are now available in the schools. These lessons are transferable to college level and might help us simultaneously develop a new generation of who would begin to disseminate a wider public understanding of science through competent teaching of science in elementary schools.

The above also underscores Arons' concern with the recent history (1950–2000) of physics education in the United States (Arons, 1993, 1997b, 1998), and his conviction that enhancing science understanding of the general population requires drastic upgrading of elementary-school education, as can only occur if, among other things, prospective teachers are far better educated in the universities of America (Goodlad, 1990, 1998, 2004; Goodlad & Keating, 1995; Hake, 2004b; Handelsman et al., 2004; NRC, 1996b).

An answer to Arons' (2000) key question: "Whence do we get the teachers with the background, understanding, and security to implement such instruction?" is crucial to the large-scale K-12 implementation of (a) the Benezet Method (Benezet, 1935/1936; Mahajan & Hake, 2000); (b) "Physics First" (Lederman, 2001); and (c) K-12 reform methods of science/math instruction generally (Hake, 2000b, 2004b; NRC, 1996a, 1996b, 1999, 2000, 2003a, 2003b, 2013). The highest priority is to alleviate the current shortage of effective science/math teachers by motivating universities to discharge their obligation to adequately educate prospective K-12 teachers. According to an NRC advisory panel headed by Melvin George (NRC, 1996b):

> Many faculty in SME&T … (Science, Math, Engineering, and Technology) … at the postsecondary level continue to blame the schools for sending underprepared students to them. But, increasingly, the higher education community has come to recognize the fact that teachers and principals in the K-12 system are all people who have been educated at the undergraduate level, *mostly in situations in which SME&T programs have not taken seriously enough their vital part of the responsibility for the quality of America's teachers*" …. [[(My *italics*)]]

Thus the improvement of SMET education at the undergraduate level along lines advised by Arons appears especially crucial to increasing the number of effective K-12 teachers, since prospective teachers derive little conceptual understanding from traditional introductory science courses and then tend to teach as they were taught with similar negative results. The upgrading of K-12 education might serve to raise the appallingly low level of science literacy among the general population (Hake, 2000) and thus increase our chances of solving the monumental science-intensive problems (political, economic, social, and environmental) that beset us.

ACKNOWLEDGMENTS

This is an updated and improved version of an unpublished manuscript "The Arons Advocated Method" (Hake, 2004a). This work was partially supported by NSF Grant DUE/MDR-9253965. The author thanks Jerry Epstein, Ibrahim Halloun, and David Hestenes for valuable comments; and psychologist Robert Grossman http://bit.ly/1zGDvk0, whose work in applying some of Arons' insights to introductory psychology instruction stimulated the writing of this article.

REFERENCES

Insert the following text: All URLs accessed on April 3, 2015, most shortened by http://bit.ly/

Anderson, J. R. (1987). Skill acquisition: Compilation of weak-method problem solutions. *Psychological Review*, *94*, 192–210. Retrieved from http://bit.ly/1AWYmAp

Anderson, J. R. (1993). *Rules of the mind*. New York, NY: Psychology Press. Retrieved from http://bit.ly/1D0VsPN

Anderson, J. R. (2015). *Cognitive psychology and its implications* (8th ed.). New York, NY: Worth. Retrieved from http://bit.ly/1ucMULz First published by Freeman in 1980. See also *Cognitive Psychology*, 4th ed. (Medin, Ross, & Markman, 2001)] and *Cognitive Psychology*, 6th ed. (Sternberg, 2011).

Arons, A. B. (1959). Structure, methods, and objectives of the required freshman calculus-physics course at amherst college. *American Journal of Physics*, *27*(9), 658–666. Retrieved from http://bit.ly/HBtxVs

Arons, A. B. (1965). *Development of concepts of physics*. Reading, MA: Addison Wesley.

Arons, A. B. (1973). Toward wider public understanding of science. *American Journal of Physics*, *41*(6), 769–782. Retrieved from http://bit.ly/1datg0Z

Arons, A. B. (1974). Toward wider public understanding of science: Addendum. *American Journal of Physics*, *42*(2), 157–158. Retrieved from http://bit.ly/1w1cc6g

Arons, A. B. (1976). Cultivating the capacity for formal reasoning: Objectives and procedures in an introductory physical science course. *American Journal of Physics*, *44*(9), 834–838. Retrieved from http://bit.ly/1BaY240

Arons, A. B. (1977). *The various language: An inquiry approach to the physical sciences; With teacher's guide*. New York, NY: Oxford University Press. For a perceptive review see Walker (1978).

Arons, A. B. (1983). Student patterns of thinking and reasoning. Pt. 1. *Physics Teacher*, *21*(9), 576–581. Retrieved from http://bit.ly/1zVaJMt

Arons, A. B. (1984a). Student patterns of thinking and reasoning. Pt. 2. *Physics Teacher*, *22*(1), 21–26. Retrieved from http://bit.ly/12zu7DN

Arons, A. B. (1984b). Student patterns of thinking and reasoning. Pt. 3. *Physics Teacher*, *22*(2), 88–93. Retrieved from http://bit.ly/12zuHRW

Arons, A. B. (1990). *A guide to introductory physics teaching*. New York, NY: Wiley. Reprinted with minor updates in Arons (1997a). The present article sites page numbers for both the 1990 and 1997a editions. For reviews see Holbrow (1990) and Knight (1995). For comments on Holbrow's review see Hake (1991b).

Arons, A. B. (1993). Uses of the past: Reflections on United States physics curriculum development, 1955 to 1990. *Interchange*, *24*(1–2), 105–128.
"An abstract, online at http://bit.ly/19WQcQ6 reads:

> A personal assessment is made of the United States experience in physics curriculum development and teacher education over the past 35 years. It is concluded that, although substantial improvements were achieved in the development of hands-on primary and secondary school curricula, their effectiveness was in large part negated by logistic problems in implementation and deficiencies in teacher preparation. At the college/university level, existing materials tend to cover far too much material far too rapidly, and, except for small, separated pockets of promising activity, inadequate attention is being paid, in the aggregate, to students' levels of cognitive development and to the learning problems revealed in research on teaching and learning. Some directions of effort that might prove helpful are described."

Arons, A. B. (1997a). *Teaching introductory physics*. New York, NY: Wiley. Contains a slightly updated version of Arons (1990), plus "Homework and Test Questions for Introductory Physics Teaching" (Arons, 1994), plus a new monograph, *Introduction to Classical Conservation Laws*." The present article cites page numbers for both the 1990 and 1997a editions.

Arons, A. B. (1997b). Improvement of physics teaching in the heyday of the 1960's. In J. Wilson (Ed.), *Conference on the introductory physics course on the occasion of the retirement of Robert Resnick* (pp. 13–20). New York, NY: Wiley.

Arons, A. B. (1998). Research in physics education: The early years. In T. C. Koch & R. G. Fuller (Eds.), PERC 1998: Physics education research conference proceedings 1998. Retrieved from http://bit.ly/1s0gVhK

Arons, A. B. (2000). Private communication to R.R. Hake of 30 June concerning Louis Paul Benezet's (1935/1936) educational methods.

Benezet, L. P. (1935/1936). The teaching of arithmetic I, II, III: The story of an experiment. *Journal of the National Education Association, 24*(8), 241–244. (1935); 24(9), 301–303 (1935); 25(1), 7–8 (1936). Reprinted in the Humanistic Mathematics Newsletter 6: 2–14 (May 1991); available along with other Benezetia at the Benezet Centre. Retrieved from http://bit.ly/926tiM See also "Is it time for a physics counterpart of the Benezet/Berman math experiment of the 1930" (Mahajan & Hake, 2000).

Bransford, J. D., Brown, A. L., & Cocking, R. R. (Eds.). (2000). *How people learn: Brain, mind, experience, and school* (Expanded ed.). Washington, DC: National Academies Press. Retrieved from http://bit.ly/1yIq4BG

Diamond, M. C. (1996). The brain … use it or lose it. *Mindshift Connection, 1*(1). Retrieved from http://bit.ly/1zlPDaS

Donnally, B. (1973). Arnold arons: Oersted medalist for 1972. *American Journal of Physics, 41*, 768. Retrieved from http://bit.ly/1zVDdWs

Goldberg, M. F. (1995). A portrait of John Goodlad. *Educational Leadership, 52*(6), 82–85. Retrieved from http://bit.ly/1uw5gaj Goldberg wrote:

> The acknowledged leader of educational renewal, John Goodlad has been at the vortex of every wind that has blown across education since World War II – holding firm to the basic ideas of humanism and progressivism. From the 1970s, when he took a stand against the behavioral objectives movement, to the early 1990s, when he opposed "America 2000" … [[proposed by G. H. W. Bush – see http://bit.ly/1Fac6Ig]] … Goodlad has steadfastly adhered to the wisdom of Alfred North Whitehead and John Dewey.

Goodlad, J. I. (1990). *Teachers for our nation's schools*. San Francisco, CA: Jossey-Bass.

Goodlad, J. I. (1998). *Educational renewal: Better teachers, better school*. San Francisco, CA: Jossey-Bass.

Goodlad, J. I. (2004). *A place called school: Twentieth anniversary edition* (2nd ed.). New York, NY: McGraw-Hill. "The description reads:

> First published 20 years ago, A Place Called School is the revolutionary account of the largest on-scene study of U.S. schools ever conducted. Carried on over four years, trained investigators entered more than 1,000 classrooms nationwide to talk to teachers, students, administrators, parents, and other community members. The result is this report. Written by one of the nation's most astute and experienced educators, Goodlad's message of optimism and his agenda for

improvement have only grown in importance since the book's original publication."

See also: *A Portrait of John Goodlad* (Goldberg, 1995); the Wikipedia entry at http://en.wikipedia.org/wiki/John_Goodlad; (c) the video interviews at http://bit.ly/1shf4uP; and (d) *Remembering Education Research Pioneer John I. Goodlad* (Sparks, 2014).

Goodlad, J. I., & Keating, P. (1995). *Access to knowledge: The continuing agenda for our nation's schools, foreword by D.M. Stewart, College board.*

Grossman, R. W. (2005). Revealing hidden transformations: Making science more learnable. *College Teaching*, *53*(1), 33–40. Retrieved from http://bit.ly/1D0EvFd See also Grosssman (2009b). Grossman wrote:

> [With regard to "hidden transformations"] Craig Nelson (1989), suggested I read a book by Arnold Arons, A Guide to Introductory Physics Teaching (1990). The content of this book is also contained in Arons's, Teaching Introductory Physics (1997a), along with several other valuable sections. As I read Arons's list of key underpinnings necessary for students to understand any introductory physics course, I could see he that he broke up the learning into small enough steps so that students were explicitly taught knowledge transformations that were so often skipped over by other physics instructors. Similarities between what he discussed and the knowledge transformations that I expected my psychology students to make was apparent. If these ideas made some sense of the cognitive problems that students encountered in two such widely differing fields as psychology and physics, I thought other disciplines might be able to use these as a heuristic aid in finding how knowledge transformations applied to their teaching as well. It may even be that facility with these knowledge transformations is a significant part of the difference between a novice and an expert in many fields (Anderson 1987, 1993; Kurfiss 1988; Medin, Ross, and Markman 2001).

Grossman, R. W. (2009a). Structures for facilitating student reflection. *College Teaching*, *57*(1), 15–22. Retrieved from http://bit.ly/1bV3XQN The abstract reads:

> The goal of this article is to describe a continuum of levels of reflection. It briefly focuses on Deanna Kuhn's research into the development of scientific thinking and Robert Kegan's Object-Subject Theory of Development applied to the problems of inspiring students to be able to reflect. Assignments for improving students' ability to reflect are presented. Examples of student reflections are provided. These may be especially helpful for faculty in a wide range of courses.

Grosssman. (2009b). In a pre-publication version, online on Grossman's. Retrieved from http://bit.ly/1zGDvk0 (4th link down on list to the left).

Hake, R. R. (1987). Promoting student crossover to the Newtonian world. *American Journal of Physics*, *55*(10), 878–884. Retrieved from http://bit.ly/a6vc3H. See also Hake (1992).

Hake, R. R. (1991a). My conversion to the arons-advocated method of science education. *Teaching Education*, *3*(2), 109–111. Retrieved from http://bit.ly/h2Ya0l

Hake, R. R. (1991b). Socratic Pedagogy in introductory physics, Letter to the editor in response to Holbrow's (1990) comment: "... is its program [Arons' *Guide to Introductory Physics Teaching*] for improving teaching of introductory physics feasible?". *Physics Today*, *44*(9), 110. Retrieved from http://bit.ly/1snYN7b Therein I wrote

Our four years experience [Hake (1987); Tobias & Hake (1988)] in using the method Arons advocates shows that not only is it feasible, but that it is extremely successful in promoting students' conceptual understanding of Newtonian mechanics as measured by the Halloun-Hestenes (1985a,b) test.

Hake, R. R. (1992). Socratic pedagogy in the introductory Physics Lab., *Physics Teacher*, *30*, 546–552. Retrieved from http://bit.ly/9tSTdB

Hake, R. R. (1998a). Interactive-engagement vs traditional methods: A six-thousand-student survey of mechanics test data for introductory physics courses. *American Journal of Physics*, *66*, 64–74. Retrieved from http://%bit.ly/d16ne6

Hake, R. R. (1998b). *Interactive-engagement methods in introductory mechanics courses*. Retrieved from http://bit.ly/aH2JQN A crucial companion paper to Hake (1998a): average pre/post test scores, standard deviations, instructional methods, materials used, institutions, and instructors for each of the survey courses of Hake (1998a) are tabulated and referenced. In addition the paper includes: (a) case histories for the seven IE courses of Hake (1998a) whose effectiveness as gauged by pre-to-post test gains was close to those of T courses, (b) advice for implementing IE methods, and (c) suggestions for further research.

Hake, R. R. (2000). The general population's ignorance of science related societal issues: A challenge for the university. *AAPT Announcer*, *30*(2), 105. Retrieved from http://bit.ly/9LxKOL Based on an earlier libretto with the leitmotiv: "The road to U.S. science literacy begins with effective university science courses for pre-college teachers." The opera dramatizes the fact that the failure of universities throughout the universe to properly educate pre-college teachers is responsible for our failure to observe any signs of either terrestrial or extraterrestrial intelligence.

Hake, R. R. (2002a). Socratic Dialogue Inducing Laboratory Workshop. *Proceedings of the UNESCO-ASPEN Workshop on Active Learning in Physics*, University of Peradeniya, Sri Lanka, 2–4 December. Retrieved from http://bit.ly/qMo1ES and as ref. 28 at http://bit.ly/a6M5y0

Hake, R. R. (2002b). Lessons from the physics education reform effort. *Conservation Ecology*, *5*(2), 28. Retrieved from http://bit.ly/aL87VT. This online journal is now called *Ecology and Society* http://bit.ly/12WX8ZQ

Hake, R. R. (2002c). Physics First: Precursor to Science/Math Literacy for All? *APS Forum on Education Newsletter*, Summer. Retrieved from http://bit.ly/JzlVrII

Hake, R. R. (2003). Socratic Dialogue Inducing (SDI) Labs website. Retrieved from http://bit.ly/9nGd3M Contains a brief description of SDI Labs with pertinent references and 10 SDI Lab Manuals that may be freely downloaded.

Hake, R. R. (2004a). *The arons advocated method*, unpublished manuscript. Retrieved from http://bit.ly/boeQQt

Hake, R. R. (2004b). Direct Science Instruction Suffers a Setback in California – Or Does It? Contributed to the 129th National AAPT meeting in Sacramento, CA, 1–5 August. Retrieved from http://bit.ly/aWsazm and as ref. 33 at http://bit.ly/a6M5y0. The PowerPoint slides shown at the Sacramento meeting are available at http://bit.ly/d40ksO

Hake, R. R. (2012). Helping students to think like scientists in socratic dialogue inducing labs. *Physics Teacher*, *50*(1), 48–52. Retrieved from http://bit.ly/wLy3En. A version identical to the Phys. Teach. article except for minor formatting changes and the addition of a few hot-linked URLs is. Retrieved from http://bit.ly/x5ruYF

Handelsman, J. (2006). *Scientific teaching*. New York, NY: W.H. Freeman.

Handelsman, J., Ebert-May, D., Beichner, R., Bruns, P., Chang, A., DeHaan, R., & Wood, W. B. (2004). Scientific teaching. *Science*, *304*(5670), 521–522. Retrieved from http://bit.ly/ILrHBK. See also the supporting material. Retrieved from http://bit.ly/eOCJmo. See also Handlesman (2006) and Miller, Pfund, Pribbenow, and Handelsman (2008).

Haskell, P. (2001). *Arnold Arons Obituary*. Retrieved from http://%bit.ly/1xmI9Gj. Peter Haskell http://www.peterhaskell.com/ is a son-in-law of Arnold Arons.

Hoellwarth, C., Moelter, M. J., & Knight, R. D. (2005). A direct comparison of conceptual learning and problem solving ability in traditional and studio style classrooms. *American Journal of Physics*, *73*(5), 459–463. Retrieved from http://bit.ly/156hHhF

Howe, K. (2001). Qualitative educational research: The philosophical issues. In V. Richardson (Ed.), *Handbook of research on teaching* (4th ed., pp. 201–208). Washington, DC: American Educational Research Association. Retrieved from http://bit.ly/1DDqQ7j

Howe, K., & Eisenhart, M. (1990). Standards for qualitative (and Quantitative) research; A prolegomenon. *Educational Researcher*, *19*(4), 2–9. Retrieved from http://%bit.ly/1uQ74Mz

Knight, R. (1995). Review of a guide to *Introductory physics teaching (Arons, 1990), APS Forum on Education Newsletter*, August. Retrieved from http://bit.ly/1wbwbxM

Kurfiss, J. G. (1988). Critical thinking: Theory, research, practice, and possibilities, *ASHE-ERIC higher education report no. 2*. Association for the Study of Higher Education. Retrieved from http://1.usa.gov/1woAV1y

Lawson, A. E. (2006). Points of view: On the implications of neuroscience research for science teaching and learning: Are there any? *CBE-Life Sciences Education*, *5*, 111–117. Retrieved from http://bit.ly/1xnhOrC

Leamnson, R. (2000). Learning as biological brain change. *Change Magazine*, *432*(6), 34–40. Retrieved from http://bit.ly/1yOaEZ2. An excerpt is online at http://bit.ly/1wQn0EU

Lederman, L. (2001). Revolution in science education: Put physics first. *Physics Today*, *54*(9), 11–12. Retrieved from http://%xfin.tv/1zq3DjO

Mahajan, S., & Hake, R. R. (2000). Is it time for a physics counterpart of the Benezet/Berman math experiment of the 1930's? *Physics education research conference 2000: Teacher education*. Retrieved from http://bit.ly/1hGDBjs and as article #6, at the Benezet Centre http://www.bit.ly/926tiM

Matthews, M. R. (2001). Obituary: Arnold Arons. Retrieved from http://bit.ly/1ubZsDP (scroll to the bottom). Michael Matthews is the editor of *Science and education* http://bit.ly/12jF1gP

McDermott, L. C., & Redish, E. F. (1999). RL-PER1: Resource letter on physics education research. *American Journal of Physics*, *67*(9), 755–767. Retrieved from http://bit.ly/10hjC4R

McDermott, L. C., Wilson, K. G., & Jossem, E. L. (2001). Arnold arons obituary. *Physics Today*, *54*(9), 76–77. Retrieved from http://bit.ly/1wt0yBC

Medin, D. L., Ross, B. H. & Markman, A. B. (2001). *Cognitive psychology* (4th ed.). New York, NY: Wiley. See also *Cognitive Psychology* (8th ed.) (Anderson, 2015) and *Cognitive Psychology* (6th ed.) (Sternberg, 2011).

Meltzer, D. E., & Thornton, R. K. (2012). Resource letter ALIP-1: Active-learning instruction in physics. *American Journal of Physics*, *80*(6), 478–496. Retrieved from http://bit.ly/O35gtB

Miller, S., Pfund, C., Pribbenow, C. M., & Handelsman, J. (2008). Scientific teaching in practice. *Science*, *28*(5906), 1329–1330. Retrieved from http://bit.ly/Jx3TEW, http://bit.ly/JlZYGy. Supporting online material available at http://bit.ly/IKOSfD

Minstrell, J. (1981). Talks with great teachers: Arnold arons. *Physics Teacher*, *19*(8), 520–526. Retrieved from http://bit.ly/1udT26i

NRC (1996a). *National science education standards*. Washington, DC: National Academy Press. Retrieved from http://bit.ly/OySndK

NRC. (1996b). *Shaping the future: New expectations for undergraduate education in science, mathematics, engineering, and technology*, advisory committee chaired by Melvin George. Retrieved from http://1.usa.gov/fs0cOU. See also NSF (1994).

NRC. (1999). *Transforming undergraduate education in science, mathematics, engineering, and technology*, National Research Council, Committee on Undergraduate Science Education, National Academy Press. Retrieved from http://bit.ly/1lZ4kpQ

NRC. (2000). *Inquiry and the national science education standards*: A guide for teaching and learning. Washington, DC: National Academy Press. Retrieved from http://bit.ly/1FV8FI0 The description reads, in part:

> Inquiry and the National Science Education Standards is the book that educators have been waiting for – a practical guide to teaching inquiry and teaching through inquiry, as recommended by the National Science Education Standards. This will be an important resource for educators who must help school boards, parents, and teachers understand "why we can't teach the way we used to." "Inquiry" refers to the diverse ways in which scientists study the natural world and in which students grasp science knowledge and the methods by which that knowledge is produced. This book explains and illustrates how inquiry helps students learn science content, master how to do science, and understand the nature of science.

See especially biologist Bruce Alberts' foreword: "A Scientists Perspective on Inquiry" for a definition of "inquiry teaching":

> Teaching science through inquiry allows students to conceptualize a question and then seek possible explanations that respond to that question … Inquiry is in part a state of mind – that of inquisitiveness.

NRC. (2003a). In M. A. Fox & N. Hackerman (Eds.), *Evaluating and improving undergraduate teaching in science and technology, engineering, and mathematics*. NRC Committee on Undergraduate Science Education. Washington, DC: National Academy Press. Retrieved from http://bit.ly/1BI7USW

NRC. (2003b). In McCray, DeHaan, & Schuck (Eds.). *Improving undergraduate instruction in science, technology, engineering, and mathematics: Report of a workshop*. NRC Committee on Undergraduate STEM Instruction, National Research Council. Washington, DC: National Academy Press. Retrieved from http://bit.ly/y6cFAG

NRC. (2013). *Adapting to a changing world – Challenges and opportunities in undergraduate physics education*. Washington, DC: National Academy Press. Retrieved from http://bit.ly/126os6j Here the NRC finally overcomes the failure of previous reports NRC (2003a, 2003b) to recognize the importance of measuring average pre-to-posttest normalized learning gains on concept inventories http://bit.ly/dARkDY. The report states:

> Hake's (1998a) seminal report http://bit.ly/9484DG on the effectiveness of interactive engagement methods remains an important contribution to undergraduate

physics education. The article presents results from the Mechanics Diagnostic (MD) [Halloun & Hestenes (1985) at http://bit.ly/fDdJHm] and the Force Concept Inventory (FCI) [Hestenes et al. (1992) at http://bit.ly/foWmEb], given before and after instruction on Newtonian mechanics in a variety of courses taught using different approaches … the conclusion, that more effective instructional approaches involve active learning, has been supported by many other studies using different methodology [Meltzer and Thornton (2012) at http://bit.ly/O35gtB and Hoellwarth, Moelter, and Knight (2005) at http://bit.ly/156hHhF].

NSF. (1994). *Project impact: Dissemminating innovation in undergraduate education*. Retrieved from http://bit.ly/1GtB9Jf. Scroll down 0.85 of the way to the end to see details of various physics education reform projects funded by the NSF as of 1994.

Shavelson, R. J., & Towne, L. (2002). *Scientific research in education*. Washington, DC: National Academy Press. Retrieved from http://bit.ly/1DEdnMF

Sparks, S. D. (2014). Remembering Education Research Pioneer John I. Goodlad, *Education Week*, 03 December. Retrieved from http://bit.ly/1yX1Xzi Sparks wrote:

Mr. Goodlad's education career ran the gamut: He wrote or edited more than a dozen books, founded the Seattle based Institute for Educational Inquiry, co-founded the Center for Educational Renewal, worked as a professor emeritus in the University of Washington's College of Education, and served as president of the American Education Research Association from 1967 through 1968.

Sternberg, R. J. (2011). *Cognitive psychology* (6th ed.). San Diego, CA: Harcourt. See also *Cognitive Psychology,* 4th ed. (Medin, Ross, & Markman, 2001)] and *Cognitive Psychology and its Implications,* 8th ed. (Anderson, 2015)].

Tobias, S., & Hake, R. R. (1988). Professors as physics students: What can they teach us? *American Journal of Physics*, *56*(9), 786–794. Retrieved from http://bit.ly/eaqFzU

UMPERG. (2015). *University of Maryland physics education research group: Physics education research papers on the web*. Retrieved from http://bit.ly/1A8NePX

Woods Hole Oceanographic Institution. (2001). *In memorium*, Arnold Arons. Retrieved from http://bit.ly/1ubKHRj

Zull, J. E. (2002). *The art of changing the brain: Enriching the practice of teaching by exploring the biology of learning*. Sterling, VA: Stylus Press. Retrieved from http://bit.ly/1atqCTW

PART II
PRACTICES AND STRATEGIES

MIT'S INTRODUCTION TO BIOLOGICAL ENGINEERING: A LONGITUDINAL STUDY OF A FRESHMAN INQUIRY-BASED CLASS

Natalie Kuldell and Rudolph Mitchell

ABSTRACT

We describe an introductory class in biological engineering that uses project-based and mentored inquiry to create a supportive, exciting, and effective learning environment. Freshman students at MIT work in small teams and with senior MIT students to design a biotechnology that addresses a real-world challenge of their choosing. Students gain familiarity with the tools and vocabulary for biodesign first through some hands-on experiences with synthetic biological systems and later by working in teams to define, present and then refine their ideas. A multi-year study of the class experience and impact included postsurveys and semistructured interviews of two freshman cohorts and a retrospective survey of three freshman cohorts. Data support the claim that students perceive academic gains through their project-based classroom

Inquiry-Based Learning for Science, Technology, Engineering, and Math (STEM) Programs: A Conceptual and Practical Resource for Educators
Innovations in Higher Education Teaching and Learning, Volume 4, 75–92

ISSN: 2055-3641/doi:10.1108/S2055-364120150000004006

experience. Freshmen reported they are better able to understand content in some of their other MIT courses, are better able to read scientific articles, and now think differently about biology. Moreover, they indicated the class was valuable in learning technical content and synthetic biology. We find this project-based class helps students make meaningful connections to scientific ideas, to personal goals and to a vision of their future selves.

Early in their academic careers, freshmen students at Massachusetts Institute of Technology (MIT) are hard pressed to identify anything, anything at all, that they don't like about MIT. When explicitly asked by the Dean of Undergraduate Education about disappointments or let-downs, a look of confusion comes to their faces. "What's not to love?" is the implied or explicit answer they give. They feel that the people are great, the excitement is palpable, and they can't wait for classes to start. Yet poll those same students a year later when they return to campus for the sophomore year, and we find they have a slew of misgivings and a ready list of discontent. And while some change in attitude is reasonable to expect, given the rigor of MIT's freshman curriculum and the personal growth that most students experience in college, there is clearly a culture shock for most students and a perceptible deflating of the spirit in many.

All of MIT's freshmen take General Institute Requirements in their freshman year including a year of Physics and Calculus, and a term of Chemistry and Biology. These classes usually enroll several hundred students in the lecture portions and so are often far larger than the students experienced in high school. The classes are also considerably more academically challenging. On MIT's admission blog (MIT Admissions Office, 2008), one student who struggled in an accelerated freshman Physics class wrote,

> You know, people say that physics is only learned through practice, but I quickly realized that I had an issue. The class was moving too fast for me to practice. The lectures didn't teach ANYTHING. They consisted entirely of PowerPoint presentations that talked about theories, general concepts, and information that never actually taught me how to do physics.

The disconnect between theory and practice is evident in this student's description of his experience, and many students are surprised by the dramatically elevated expectations as they move from high school to college. The happy ending for this freshman's story is that he switched to a slower-paced section of physics and ultimately earned a "B" in the class.

But believing that his story is representative of the culture shock most freshmen experience at MIT during their first year, MIT's Task Force on the Undergraduate Educational Commons wanted to improve the freshman experience, and they recommended our Institute experiment with project-based (PB) classes designed for freshmen as a way to sustain their motivation and enthusiasm for learning (Healey, 2005). Informed by some successful PB-classes MIT teaches for upperclassmen and with financial support from the d'Arbeloff Fund for Excellence in Education, six new PB-classes for freshmen were taught for the first time in the 2006–2007 academic year (Freeman, Cooper, & Lucas, 2007). These were "add-on" classes to the standard collection that freshman enroll in. The home departments offering these classes ranged widely, from Mechanical Engineering to Architecture and Urban Planning. However the PB-classes had unifying goals that included design-motivated cross-disciplinary learning, synthesis of ideas and techniques, and the use of real-world problems to drive disciplinary knowledge. Students in these classes had an opportunity to develop solutions that met personally meaningful challenges. The solutions were devised through their own informed decision-making process rather than by their work towards a single-solution that had been prescribed in advance.

Coincident with these initial offerings of freshman PB-classes, MIT's Department of Biological Engineering was established. Offering the first new undergraduate major at MIT in nearly 30 years, the Department of Biological Engineering was undertaking a number of educational experiments of its own. In particular, the new Department designed several classes whose the explicit goal was to define Biological Engineering as an engineering discipline and to distinguish this new major from existing programs in Biomedical Engineering and in Biology. The tagline used by our new department, "creating biological technologies, from discovery to design," was broad enough to incorporate many ideas and approaches. It was also so broad that it articulated little about our educational mission or approach. To help freshmen students understand MIT's new undergraduate major and this new field, Professor Drew Endy and I (NK) taught a freshman PB-class starting in the Spring of 2008. We named the class "Introduction to Biological Engineering Design," and it was given the MIT course number of 20.020 (Kuldell, 2013b), which is read aloud as "twenty-twenty."

We liked this number a lot since it reflected our hope that the class would help students perfectly envision the roll that Biological Engineering could play in their MIT education, in their future careers and in the world. Though Professor Endy left MIT in 2009 to teach at Stanford University, the class has been taught for 12–20 students by one of us (NK) and evaluated by another of us (RM) each Spring since 2008.

20.020 gives students the opportunity to design a biotechnology that is responsive to a real world need. Over the 14-week term, we guide the class through different "phases" transitioning from whole group activities (labs, videos, role-playing etc.) and teacher-led discussions in phase 1, to work in small teams, student presentations and consultations with technical experts in phase 2. We mentor the freshmen students in multiple ways while they discover their interests and as they take increasingly more ownership and responsibility for the project they've devised. Meaningful learning comes through this process of authentic, mentored inquiry (Donovan & Bransford, 2005; Freeman et al., 2014). Inquiry-based classrooms avoid the "mental slavery" that John Dewey cautioned against nearly a century ago (Dewey, 1938/1997), and directly support student interest and engagement (Boyer, 1998). Inquiry-based classrooms are inherently authentic, interdisciplinary, and reflective because students ask their own questions (Cummins, Green, & Elliott, 2004; Luckie et al., 2012), a skill that is viewed by some as the single most valuable thing we can teach them (Rothstein & Santana, 2011).

CLASS STRUCTURE

Phase 1: Provide a Conceptual Framework for Synthetic Biology

At the start of the term, 20.020 is taught through short interactive lectures, and with hands-on activities intended to give students direct experience with engineered living systems. These lectures and activities establish the basic techniques, strategies, and vocabulary for biological engineering. For example, the first few lecture hours emphasize an abstraction hierarchy that is used in a subfield of biological engineering called synthetic biology. The abstraction breaks complex systems into manageable and distinct levels of detail. At the highest level, called "Systems," a plain language description of the biotechnology is required. For example, "I want a cell that can detect arsenic and turn red when it detects unsafe levels of the contaminant" is one system-level description we use in the class that meets an environmental grand challenge. At the other end of the abstraction hierarchy is the DNA sequence that encodes the genetic parts to build the living system. The students are shown where such sequences can be found on the Internet, and they come to appreciate how modern scientific understanding and their own grasp of it enable the engineering of living systems. We return to this aspect of the class later in this chapter because, according to our assessments, its impact on their experience is tremendous.

The engineering strategies and vocabulary the students hear about in lecture are reinforced by direct experimentation. We provide some prototypical examples of engineered living systems for them to measure, manipulate, and test. These hands-on activities include smell tests of bacteria engineered to smell like bananas, and introduction of DNA into bacteria to turn the cells purple or green. There are activities in which students use computers to model and simulate the behavior of unnatural DNA circuits. We also set up roll-playing activities in which they anticipate public reaction to biological engineering as an emerging technology. The variety and charisma of these activities really "hook" the students on the potential of engineering the living world, as well as raise their awareness of science and engineering in society. During interviews of the students conducted by one of the authors (RM), we hear comments that echo this student's (caseFL) comment, "When you are doing hands-on stuff, everyone is pretty much excited. I think that is just the nature of the work. And everyone is like, 'Oh, this is really cool to see as the end result'." The activities requiring microbes are sourced from our bioengineering teaching laboratory that is physically near the PB-classroom, but all the activities can be carried out in a traditional (non-laboratory) classroom with impervious and paper-covered tables to keep the reagents, and after Biosafety Level 1 training of the students by our office of Environmental Health and Safety.

Phase 2: Design a Biotechnology

The activities and vocabulary that are central to "Phase I" of 20.020 provide a platform and common experience that students draw on as they move into the design phase of the class. One student, caseTL, noted how, "We started by designing our general idea of the system, an abstract version of it. Natalie had taught us about black-boxing things. [We first identified] what we wanted the system to do. Then we focused on how we would actually get it to do that."

Intentionally, the Phase I lectures and activities apply equally well to any of the challenge areas that students are considering for their own projects. Topic areas such as "food and energy," "health," or "environment," are deliberately kept broad, and students have the freedom to choose which topic area to work in. We could have chosen to focus the class so that all students address a common goal, such as clean water one year, and sustainable fuel the next. However, since we wanted students to feel genuine investment in their project, we allow them to choose the challenge area and to set their own goal. Furthermore, because they are freed from the obligation of implementing their final designs in a lab, they are happily unencumbered by the voices that say "it won't work" or "that can't be done here." One faculty member emailed, after hearing about a project idea that aimed at remylenating neurons, to say, "Sounds exciting. I would argue that not knowing what is possible in the brain, however, can be useful for the synthetic biologist – maybe something new can be created, from scratch!"

Project Management

If shepherding a classroom full of open-ended student-selected projects seems daunting, there are at least three aspects that we've implemented in our PB-class that we find make the effort manageable. The first of these aspects is class size, admittedly the one aspect that teachers cannot easily control. Scheduling conflicts, interest level and who knows what else typically keeps 20.020 enrollment to two dozen students each spring, with approximately three-quarters of them ultimately completing the class. Another aspect that keeps our open-ended PB-class manageable is that students generally group themselves into a small number of project areas, five or fewer, with anywhere from two to seven students in each group. Why this happens isn't clear.

The aspect of our class that has the greatest impact on its management is an unusual overlay we've designed in which a second class runs as a companion class with the freshman PB-class, a seminar course called "20.385: Current Research in Synthetic Biology" (Kuldell, 2013a). This class runs with anywhere from six to 12 students who are generally juniors or seniors majoring in biological engineering at MIT. Students in this advanced class not only read and discuss the current primary literature each week, but they are also assigned as a mentor to one team of freshman, working with them during the long block of "studio time" that is set aside for the freshman to develop their project ideas. Such interactive, constructive activities have been shown to increase student learning outcomes (Menekse, Stump, Krause, & Chi, 2013; Mohamed, 2008). Indeed, our assessments have invariably indicated positive outcomes for both the freshmen and the advanced students when grouped to collaboratively develop project ideas. For example, one freshman, caseSN, noted how the mentors are helpful but not prescriptive, saying "they have seen other things that we have not, and can offer their perspectives. At the same time, they are not trying to impose little stuff on us because we actually have more raw knowledge on the topic than they do." Mentors invariably know more about the ups and downs of their team's project development and the team's interpersonal dynamics than the course instructor. One freshman student, caseNA, made the following reflection.

> Our mentors are just like friends. They are like big sister and big brother .... They help us polish our slides, to make our slides better. When we present, they are in the room and listen to our presentation. During the rehearsal, they ... help us to prepare. They pretend to be the other students and raise questions.

MILESTONES

Once the teams are formed and their mentors are assigned to them, the freshmen work towards a presentation to the class of three project ideas in their grand challenge area. These "3 ideas" presentations are given six weeks after the term begins, and they address high-level aims, not implementation ideas. For example, students explicitly discuss how large a problem they are focusing on, how much their project could help, if other technologies exist to address this challenge and if there are things that are unknown or unknowable about this challenge. In the area of "food" one year, the team's three ideas were (1) tartar controlling "Bacto-gum," (2) protein-enriched rice, and (3) cells that remove toxins from clay in places where "geophagy" (literally "earth eating") is practiced. Feedback on each team's three ideas helps them decide which of the three projects to pursue for the rest of the term. For example, the teams frequently must decide between designing a solution that makes a large impact on a relatively small problem or another solution that has a small impact on a gigantic challenge. Both outcomes are equally valued in this class. Feedback also points the teams in the direction of similarly themed projects that may exist and they are given additional articles to read if there are publications that relate to their ideas.

Within one week of their three ideas presentation, the teams are required to narrow their three ideas down to one, a decision that frequently stresses team dynamics for the first time. Students often grow attached to one of the three ideas, especially if it is the one they initially proposed to the group. This winnowing from three ideas to one can also be a first-experience with evidence-based argumentation. For example, one student, caseEE, reported the following.

> In my high school, I was one of the smarter people. And here, everyone was top of their (sic) class in high school. In high school, everyone would agree with me. They would go with what I said .... Even if I were wrong, they would agree with me .... But here, everyone has [his/her] own opinions, own ideas. And sometimes we conflict because everyone has [his/her] own ideas. Everyone is really smart here. So definitely, working in groups has been a little more challenging than it was in high school because you are not really the one in charge anymore.

To help with the dynamics at this junction of their work, the freshmen teams rely on their senior mentors and on as many published journal articles they can find and understand quickly. On rare occasions, teams can't choose and so two of the three ideas remain viable, but in most cases one idea seems the most appealing for the team to pursue at the next, more technical level. The engagement of students in this choice is evidenced by comments such as that of caseFL, "I liked the project idea that my team decided to go with. It was very easy for me to do research on it and get involved with because it was something I was passionate about."

The "Tech Spec Review" is the next milestone in the team projects, and is reached intentionally quickly, just three weeks after the three ideas presentation. The team's initial technical specification requires they provide a preliminary outline using the abstraction hierarchy that was presented in the lectures and activities earlier in the term, an initial simulation of their biotechnology using a modeling tool, and, most importantly, a list of open questions that they are trying to address. For the "food" team's three ideas listed above, they decided to pursue the "bacto-gum" project, specifying a bacteria-infused chewing gum that would detect other bacterial cells that were growing in high population numbers and in the low pH environment around teeth, and then release an anti-inflammatory agent to improve gum health. Unknowns for the team at this point were things like, "how would the bacteria in the chewing gum sense pH in the mouth," and "would the system work only in the mouth and not in other places in the body where anti-inflammatory compounds would be undesirable." Teams must commit to a "GO/NO GO" decision on their project at this point in the term, and most, including the "food" team described here, feel sufficiently confident in their ability to further define and describe their project that they give their work a "GO". The rare "NO GO" teams move to a "runner-up" project from their three ideas presentation.

The final five weeks of the term are dedicated to refining the design and specification of their projects. Lectures and hands-on activities are rare but important in this home stretch of the class because they help students polish their projects. Short lectures present tips for things like measuring reliability of different systems or for giving powerpoint presentations. Teams meet in class and often outside of class research and to develop their project idea into one that could be implemented in a lab.

In addition to the guidance provided by the team mentors, the final few weeks of the project development cycle include "consultations" with two or three experts in the team's area of study. These consultations not only "up the game" for the freshmen, who are anxious about presenting their ideas to someone whose journal or news articles they've read, but the consultations also prevent the students from putting off the final push of hard work it takes to fully refine their idea. The "bacto-gum" team, having changed their project to a yeast-based one after the Tech Spec Review, met with a yeast geneticist, with a student from Harvard Dental School, and with an immunology expert, Professor Hidde Ploegh from the Whitehead Institute. Our assessment of 20.020 as a freshman PB-experience consistently shows how important these meetings with consultants are for the students.

The range of final project ideas from this PB-class is tremendous. Freshman biodesign teams have proposed living systems for accelerating composting, stemming the red-tide algae blooms, harvesting lithium from ocean water, replacing defective genes in patients with muscular dystrophy, regulating iron levels in the blood of female athletes, and many, many more blue-sky project ideas. The ideas are based on extensively researched scientific understanding, but the teams are also granted some liberty in their designs, with imagined "black boxes" to mask the complexity of genetic functions that are, as of now, incompletely understood or as yet undiscovered. As the term ends, students leave the class wishing they could extend their work to a laboratory setting. In addition to their intellectual eagerness, students have also learned to work with each other in healthy, deadline-respectful ways and they are sad to break up their teams as the term ends.

METHODOLOGY

Since 2008, one of the authors (RM) has conducted a study of the course with the 2008, 2011, 2012, and 2013 cohorts. Each cohort consists of freshmen enrolled in the spring semester of the course. For each cohort, Dr Mitchell conducted semistructured interviews; for the '12 and '13 cohorts, a postsurvey was added. In spring, 2014, a post-postsurvey was administered to the '11, '12, and '13 cohorts. In this chapter we review results from the combined '12 and '13 cohort surveys and interview data, as well as data from the post-postsurvey. The fifteen-minute, online postsurveys explored learning activities, class atmosphere, and impact, and consisted primarily of statements for which students expressed their level of agreement by means of a 7-point Likert Scale (1 = strongly disagree and 7 = strongly agree). Survey data were analyzed using IBM SPSS Statistics, version 22 (IBM Corp., 2012).

Students from the '12 and '13 cohorts participated in the interviews that were conversational, informal, and ranged in length between 20 and 30 minutes. By means of several open-ended questions, the interviews explored students' view of the experience, its impact, and their attitude toward the experience. The interviews were audio taped, and transcriptions were made of each session. A content analysis was performed on the responses (Corbin & Strauss, 2008; Maxwell, 2012; Weiss, 1994). In this chapter, two letter codes serve to distinguish participant comments.

RESULTS

Twelve of the 2012 cohort and eighteen of the 2013 cohort completed the post survey (response rate = 63% and 90%, respectively). 30 students completed the post-postsurvey (response rate = 73%). Survey links failed to reach six students who should have been surveyed with the post-postsurvey instrument, limiting the opportunity to participate to only 41 of the 47 possible respondents.

Analysis of the survey data of the interview responses (Mitchell, 2014) supports the claim that students perceive academic gains through their PB-classroom experience. For example, in their responses to the 2012 and 2013 postsurveys, for which Tables 1 and 2 provide greater detail, freshmen reported they are better able to understand content in some of their other MIT courses (M = 6.10), are better able to read scientific articles (M = 5.97), and now think differently about biology (M = 5.90). In responding to the post-postsurvey, students indicated the class was valuable in learning technical content (M = 5.76) and synthetic biology (M = 6.03). During the interviews, several students identified intellectual connections to

Table 1. Descriptive Statistics for Combined Data from 2012 and 2013 Postsurvey Items that Relate to Team Experience and Class Atmosphere.

Items	*N*	*M* (SD)	neg%	neut%	pos%
Q4e. Working in a team was an important part of my experience in 20.020	30	6.43 (1.22)	3	3	93
Q4f. The senior mentors were an important part of my experience in 20.020	30	5.53 (1.41)	7	7	87
Q6a. I found 20.020 fun to attend	30	6.63 (0.49)	0	0	100
Q6b. I found the class exciting	30	6.53 (0.57)	0	0	100
Q6c. I found myself looking forward to class	30	6.43 (0.68)	0	0	100
Q6d. I felt inspired by this class	30	6.07 (1.14)	0	13	87
Q6e. I felt like sharing my ideas in this class	30	6.20 (0.81)	0	3	97
Q6f. I felt I could take intellectual risks in class	30	6.23 (0.68)	0	0	100
Q6h. Our team ran into challenging problems	30	6.73 (0.45)	0	0	100
Q6i. Our team successfully worked through or around challenging problems	30	6.23 (1.01)	3	0	97
Q6j. Our team successfully worked through frustrations related to designing our system	30	6.27 (0.91)	3	0	97

Note: Neg = negative, neut = neutral, pos = positive. The negative category represents the percentage of students who indicated they *strongly disagree*, *disagree*, and *slightly disagree* with statement. Neutral category represents the percentage of students who indicated they were *neutral* toward the item. The positive category represents the percentage of students who indicated they *slightly agree*, *agree*, and *strongly agree* with the statement.

Table 2. Descriptive Statistics for Combined Data from 2012 and 2013 Postsurvey Items that Relate to Impact.

Survey items	*N*	*M* (SD)	neg%	neut%	pos%
Q9f. I was better able to understand content in some of my other MIT subjects (e.g. 7.01x)	30	6.10 (0.99)	0	10	90
Q9g. I am better able to read primary scientific literature	30	6.03 (0.96)	0	10	90
Q9h. I feel that I can better explain ideas to others	30	6.17 (1.02)	0	10	90
Q11d. As a result of 20.020 I now think differently about biology	30	5.90 (0.84)	0	7	93
Q11e. As a result of 20.020, I am a better problem-solver	30	5.83 (1.23)	7	10	83
Q11f. As a result of 20.020, my confidence in defending my ideas increased	30	5.90 (1.03)	0	13	87
Q11g. As a result of 20.020, my confidence in interacting effectively as part of a group increased	30	5.97 (1.03)	3	3	93
Q11h. As a result of 20.020, I am better able to read scientific articles	30	5.97 (1.00)	3	3	93

Note: Neg = negative, neut = neutral, pos = positive. The negative category represents the percentage of students who indicated they *strongly disagree, disagree,* and *slightly disagree* with statement. Neutral category represents the percentage of students who indicated they were *neutral* toward the item. The positive category represents the percentage of students who indicated they *slightly agree*, *agree,* and *strongly agree* with the statement.

other subjects they studied at MIT, as evidenced by quotes such as caseDM's. "We are doing a project right now that is really hands-on ... soaking in knowledge that 7.013 (Introductory Biology) could not pump into me. I am doing this myself." Or that of another student, caseJC, who said "the kind of things I learned in 18.03, differential equations .... It was like, 'Oh my God, I have never seen this explained in biology classes. This is really cool.' The idea that you can engineer biology is fascinating." Students credit the class structure as well as their classmates for their meaningful learning, for example one student (caseDM) made the following comment.

> [When] you are looking at articles, understanding the material, and putting a system together with your team ... you learn a lot. I have learned more from the four members of my group than I have from the research online. They have [taught] me different things ... or maybe something that I did not hear in bio or did not grasp.

The students' appreciation for the PB-framework goes beyond their increased knowledge of technical content. Students consistently remark on the intellectual energy and ownership they feel for their projects, for their teams and for their learning. They identify several aspects of the class atmosphere that evoked this emotional impact, aspects including their feelings of safety, the classroom's emphasis on openness, and the absence of excessive amounts of stress. One hundred percent of the freshmen felt they could take intellectual risks in class ($M = 6.23$) and 97% were willing to share their ideas in class ($M = 6.20$). The inter-relatedness of these aspects of the classroom atmosphere is emphasized in caseDM's comment, "It is not like ... if you mess up this little bit, ... you are going to fail. It is laid back. We are learning at our own pace. You are learning the stuff that pertains to your project." Several students commented how the instructor (NK) made high quality work and learning the priority for students. One student, caseJT, noted, "It is her supportive guidelines, the outline for the class expectations, as well as her demeanor when she is talking to us .... It is very clear she ... wants us to learn above everything else, which we are definitely doing!" In the postsurvey responses in both 2012 and 2013, 100% of the freshmen viewed the class as exciting ($M = 6.53$) and fun ($M = 6.63$), and found themselves looking forward to the class ($M = 6.43$). For 87%, the class inspired them ($M = 6.07$). Fig. 1 compares how positively students viewed different aspects of the class which were reported in the post-postsurvey. Even a year or more later, 100% of the students who responded to the post-postsurvey indicated how much fun the class was ($M = 6.67$). One commented on the post-postsurvey instrument, "My group from

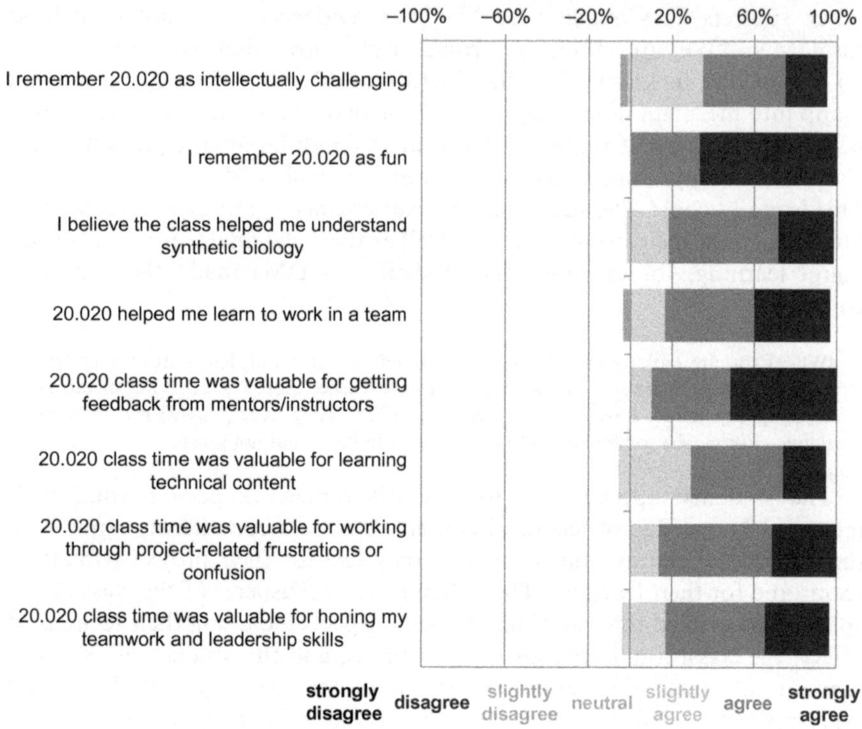

Fig. 1. Distribution of How Students Viewed the 20.010 Experience One-to-three Years after Completing the Course, Views They Reported by Means of a 7-point Likert Scale (1 = strongly disagree and 7 = strongly agree) to Statements on the Post-postsurvey. *Note*: The survey was administered in spring 2014 to the 2011, 2012, and 2013 cohorts. Thirty students completed the post-postsurvey (response rate = 73%). Survey links failed to reach six students who should have been surveyed with the post-post survey instrument, limiting the opportunity to participate to only 41 of the 47 possible respondents.

20.020 still keeps in touch and occasionally has dinner together!" and another wrote, "I really enjoyed/remember fondly all of the experiments we performed to visualize the concepts that we talked about in class!"

The emotional connection that students feel for their team and their project extends to the relationship students developed with their team mentors, the advanced students in the companion class for 20.020 who facilitate freshman teams' project development. The mentors were seen as resources for content, as able teachers, and, at times, as mediators. For example, when projects or individuals were following unproductive paths, the freshmen in 20.020 could rely on the mentors to refocus the work.

> They kept us on track. They would ask us, "Oh, how is it going? Is everything working?" Then, we would have to stop whatever we were doing and make sure that everything was working, and usually [we would] find problems. So, they were definitely useful. They did not solve problems for us, but they helped us to solve our problems. (caseAC)

With the 2012 and 2013 postsurvey data, we found 87% of the freshmen viewed the mentors as an important part of their 20.020 experience (mean = 5.53); in the post-postsurvey, 100% of the students valued their work with the team mentors (mean = 6.41).

Personal resiliency in response to setbacks and frustrations is one of the most surprising aspects of the class for the students interviewed. The safe classroom atmosphere and the support they felt from their teams and the team mentors encouraged the freshmen students to work through the challenges they faced.

> I never had to do anything like this before. So it was a new experience. It was harder than I thought it would be to [put] things together that would work to achieve our purpose, and not interfere with each other or have all sorts of horrible things happen .... Kind of frustrating ... Searching for what we needed was hard .... Once we found articles, reading through them and trying to figure out if they were actually relevant, which most of them were not, was difficult. (caseAC)

The projects posed challenges, and the challenges were compounded by the freshmen sometimes underestimating the difficulties of the design task. As one freshman, caseFL, explained,

> The [abstraction] was very helpful for me because when we first started, we just black-boxed it, like this is a device. And I was like, "Oh, this is really cool. This is the input. This is the output." I [thought,] "That is really easy." And then, when I got into it, I am like, "Okay, that is a lot more work."

Along similar lines, caseHM commented,

> In the beginning, we are working on a project, thinking "Oh, yeah, sure, that project seems pretty simple ... we can do it." But, as we read more and more, more unknowns came up, getting more complex every day .... Having [to think up] the design was not very difficult .... But finding the parts so that we can make it [work] was very challenging.

In the 2012 and 2013 postsurvey responses, 100% of the freshmen indicated they ran into challenging problems (M = 6.73), and 97% reported working successfully through or around the challenging problems, or through frustrations related to designing their systems (M = 6.23). They reported that as a result of this class, they are better problem-solvers (M = 5.83) and have gained greater confidence in defending their ideas (M = 5.90). In the post-post data, students remembered the class as intellectually challenging (M = 5.70) and valuable for working through project-related frustrations or confusion (M = 6.17) with 93% reporting it challenging and 100% reporting it valuable to working through their confusion.

Students relied on the literature and on each other to meet the challenges and frustrations arising from their work. The findings indicated that each student group moved its project forward based not on ego or personality, but based on research findings. One student, caseDM, offered this insight into the team's dynamics.

> It is kind of interesting because we do not disagree on where the project is going. [If a student] pulls up an article that says this, X, Y, and Z, I read the same thing she is reading. I am like, "Okay. That makes sense. Let us try to implement this in our project." We are not so much giving opinions. This project, we want to make it as realistic as possible. So we just go by the rule that the article we are reading is correct .... Let the research, let the articles do the talking.

Postsurvey data revealed that 93% of the freshmen reported that working in a team was an important part of their 20.020 experience (M = 6.43). In the post-postsurvey, students reported the class helped them learn how to work in teams (M = 6.10). They found class time valuable for honing their teamwork and leadership skills (M = 6.03). In every year we have taught the class, students self-identify personal improvements in teamwork as one of the most profound and satisfying aspects of the class. CaseDN's comment is typical:

> I am really close with my project group. We work really well together. Even if there were people [with] whom you did not work very well, ... [early in the semester] you learned how to work together, which I think is really important, especially in [Biological Engineering]. ... [This] is a very collaborative field and we are going to be doing a lot of group projects ... in the next three years. So, it was really important to get exposure to that.

DISCUSSION

We find that this PB-class is a supportive, exciting, and effective learning environment that helps students decide their academic and personal path.

Year after year our assessment of this PB-class reveals that the students make meaningful connections to scientific ideas, to personal goals and to a vision of their future selves.

If MIT students, who are arguably among the most STEM-oriented teenagers in the world, credit their PB-experience in 20.020 for improved understanding of technical material and as a lasting realization of their professional and interpersonal skills, then what could such a classroom experience do for students who are less inclined to study science and engineering? Asked another way: could a PB-class such as the one described here reveal for non-STEM students how creative, interdisciplinary and optimistic science and engineering can be and reveal that there is a place in this effort for them? The complex problems that engineers address have multiple solutions, and we have shown that such complexity can engage students as contributors. Open discourse around challenges in a safe environment factor into their learning, as does mentored inquiry. Fully adopting the framework we've outlined here presumes some things about class size, time, and leadership as well as resources for teaching. However there isn't any classroom that is too large, too pressed for instructional hours or too poor that couldn't adopt elements of it.

REFERENCES

Boyer, E. L. (1998). The Boyer commission on educating undergraduates in the research university, reinventing undergraduate education: A blueprint for America's research universities. Stony Brook, NY.

Corbin, J., & Strauss, A. (2008). *Basics of qualitative research*. Thousand Oaks, CA: Sage.

Cummins, R., Green, W., & Elliott, C. (2004). "Prompted" inquiry-based learning in the introductory chemistry laboratory. *Journal of Chemical Education*, *81*(2), 239–241.

Dewey, J. (1997). *Experience & education*. New York, NY: Touchstone (original work published 1938).

Donovan, M. S., & Bransford, J. D. (Eds.). (2005). *How students learn: Science in the classroom*. Washington, DC: National Academies Press.

Freeman, D. M., Cooper, E. D., & Lucas, W. A. (2007). Experimental project-based subjects: A hit with students. *MIT Faculty Newsletter*, *20*(1). Retrieved from http://web.mit.edu/fnl/volume/201/freeman.html

Freeman, S., Eddy, S. L., McDonough, M., Smith, M. K., Okoroafor, N., Jordt, H., & Wenderoth, M. P. (2014). Active learning increases student performance in science, engineering, and mathematics. *Proceedings of the National Academy of Sciences*, *111*(23), 8410–8415.

Healey, M. (2005). Linking research and teaching: Exploring disciplinary spaces and the role of inquiry-based learning. In R. Barnett (Ed.), *Reshaping the university new relationships between research, scholarship and teaching* (pp. 67–78). Maidenhead: Society for Research into Higher Education/Open University Press.

IBM Corp. Released. (2012). *IBM SPSS statistics for macintosh, version 22.0*. Armonk, NY: IBM Corp.

Kuldell, N. (2013a). *20.385 advanced topics in synthetic biology*. Retrieved from http://openwetware.org/wiki/20.20(S13):Advanced_topics

Kuldell, N. (2013b). *20.20 introduction to biological engineering*. Retrieved from http://openwetware.org/wiki/20.20(S13)

Luckie, D., Aubry, J., Marengo, B., Rivkin, A., Foos, L., & Maleszewski, J. (2012). Less teaching, more learning: 10-yr study supports increasing student learning through less coverage and more inquiry. *American Journal of Physics: Advances in Physiology Education*, *36*, 325–335.

Maxwell, J. A. (2012). *A realist approach for qualitative research*. Thousand Oaks, CA: Sage.

Menekse, M., Stump, G., Krause, S., & Chi, M. (2013). Differentiated overt learning activities for effective instruction in engineering classrooms. *Journal of Engineering Education*, *102*(3), 346–374.

MIT Admissions Office. (2008). *How hard is it really?* [Blog post], April 9. Retrieved from http://mitadmissions.org/blogs/entry/how_hard_is_it_really

Mitchell, R. (2014). *An interview study of the '12 & '13 cohorts' experience in 20.020*. Retrieved from http://hdl.handle.net/1721.1/92040

Mohamed, A. (2008). Effects of active learning variants on student performance and learning perceptions. *International Journal for the Scholarship of Teaching and Learning*, *2*(2). Retrieved from http://digitalcommons.georgiasouthern.edu/ij-sotl/vol2/iss2/11/

Rothstein, D., & Santana, L. (2011). *Make just one change teach students to ask their own questions*. Cambridge, MA: Harvard Education Press.

Weiss, R. (1994). *Learning from strangers*. New York, NY: The Free Press.

PROBLEM BASED LEARNING APPROACHES TO TEACHING CHEMISTRY

Dylan Powell Williams

ABSTRACT

This chapter will provide an overview of how Problem Based Learning (PBL) is used to support first year chemistry students at the University of Leicester, United Kingdom. The chapter will go on to provide an overview of the learning journey that we have undertaken over the last seven years by discussing the challenges we have encountered and by including details of how we have adapted our approach based on student and staff feedback and other considerations. The chapter is a follow up to a previously published article with a focus on the changes made since this initial publication (Williams, Woodward, Symons & Davies, 2010).

INTRODUCTION

Problem Based Learning (PBL) is a student-centred approach to teaching and learning (Maurer & Neuholm, 2012). The first large scale

Inquiry-Based Learning for Science, Technology, Engineering, and Math (STEM) Programs: A Conceptual and Practical Resource for Educators
Innovations in Higher Education Teaching and Learning, Volume 4, 93–112

ISSN: 2055-3641/doi:10.1108/S2055-364120150000004007

implementations of PBL were reported by the medical schools at McMaster University, Ontario (Woods, 1996). PBL has gained popularity as a teaching approach in a range of academic disciplines (Duch, Groh, & Allen, 2001) including chemistry (Belt, Enas, McCreedy, Overton, & Summerfield, 2002; Overton & Randles, 2015). The foundation of the PBL approach is an open ended problem which students tackle in small groups (Hendry, Harris, & Ryan, 2003; Newman, 2005). The open ended nature of the problems is intended to mimic real-world problems which usually have a number of different viable solutions (Savery & Duffy, 1995). PBL problems are often based on real research questions and can be used to help emphasise the interdisciplinary nature of much contemporary research (Wood & Mack, 2001). PBL can provide new opportunities for collaborative undergraduate research by bringing together students with different disciplinary specialisms to solve interdisciplinary research questions.

The PBL approach also offers a range of new opportunities to integrate authentic assessment practice into a higher education programme (e.g. students can be placed in the role of a scientific advisory team reporting back to a government committee by giving an oral presentation or by writing a formal report) (Raine & Symons, 2005). This also allows the integration of transferable skills training within the scientific curriculum of a physical science degree course.

There are a number of recognised limitations of the PBL approach including the fact it can be resource intensive (Heycox & Bolzan, 1991) and can be unsuccessful if there is a lack of high-level staff engagement (Glew, 2003).

THE RATIONALE FOR CURRICULUM CHANGE

It was decided that a new approach to teaching and learning in year one of the chemistry degree course at the University of Leicester was necessary in order to address the varied learning requirements of a heterogeneous student cohort (i.e. students with a range of different educational and social backgrounds) (Williams, Woodward, Symons, & Davies, 2010). The recent increase in student recruitment from previously underrepresented demographics (attributed to a recent increase in widening participation activities), for example students from working class backgrounds and some ethnic minorities, has led to a change in the motivation of first year science students to study the subject (Browitt & Walker, 2007). The department introduced PBL approaches in order to increase student motivation towards the subject (Barrows, 1986; Bredderman, 1983; Blumenfeld et al., 1991), to encourage a shift away from strategic learning approaches, to reduce compartmentalisation of chemical concepts (Andraos, 1999) and to address the lack of transferable skills training that was present in many UK chemistry degree courses at the start of the 21st century.

STAFF BUY-IN

The introduction of PBL approaches to an existing curriculum can be a challenge for all of those directly involved, for example staff members leading the curriculum change, students taking the course as well as student partners, and for those indirectly involved, for example staff who teach on other modules. In order to facilitate this process at the University of Leicester, a departmental champion introduced colleagues to the approach and helped prepare them for the potential challenges that lay ahead. The departmental champion brought in a consultant (from the University's Interdisciplinary Science degree programme) with experience of the approach to help oversee the development and evaluation of a series of new PBL problems which were to be used in the first year of the undergraduate chemistry programme.

IMPLEMENTATION APPROACH

It was decided that PBL would be integrated into an existing module and would form one component of the module alongside 'conventional' teaching approaches such as lectures and small group tutorials. It was hoped that the PBL part of the course would help generate an enthusiasm for chemistry amongst the students and encourage them to do further reading beyond what had been presented in the lectures. It was decided that PBL would be introduced at the start of first year in order to maximise the potential social and motivational benefits to students new to the course and to address the requirements of students at the start of their studies.

Departmental staff were notified of the curriculum change at a series of staff meetings throughout the 2006/2007 academic year. The bulk of the curriculum redevelopment work was planned to take place during summer 2007. The curriculum redevelopment process was facilitated by the appointment of a full time educational developer (the author of this chapter) whose role was to research other similar implementations and to lead the development, implementation and evaluation of a series of PBL resources.

CONSULTATION

The Centre for Interdisciplinary Science at the University of Leicester has been offering undergraduate degree courses in Interdisciplinary Science since 2003 (Raine & Symons, 2005). The Interdisciplinary Science programme is built around a backbone of PBL focused modules which are based on research themes which bring together the different physical and biological science disciplines. The Interdisciplinary Science degree course had already developed a PBL problem for a module which had similar aims and intended learning outcomes to the 'Chemical Principles' first year chemistry module. A consultant from the Centre helped the Department of Chemistry modify the existing resource to meet the requirements of the chemistry course.

REDESIGNING CONTENT

In order to redevelop the existing PBL resources, the intended learning outcomes of the resources were analysed in order to determine the extent of alignment with the existing intended learning outcomes of the 'host' chemistry module. Some of the areas of content were modified in order to be suitable for the year one chemistry module. This exercise demonstrated that it is possible to use existing resources as a way of introducing PBL to a course but that careful consideration of the suitability of all aspects of the resource from the 'donor' programme must be considered and, modifications may need to be made in order for the resources to be suitable for the 'host' programme.

REDESIGNING THE STUDENT LEARNING PROCESS

The initial PBL implementation involved dividing students into groups of approximately six (partly influenced by examples of good practice elsewhere (e.g. Newman, 2005) and partly dictated by timetable requirements) who were required to attend two one hour contact (facilitation) sessions per week. Students were assigned to groups by staff. Staff tried to ensure there was a suitable gender balance in each group but group composition was not informed by previous academic performance. The PBL part of the module was divided into a series of short problems which covered a ten week period. Students were given access to the problems at least four days ahead of their facilitation session via the virtual learning environment (Blackboard).

Students may need to use a different set of skills to solve PBL problems to those they use to solve conventional 'closed' problems (e.g. end of chapter questions). In order to support the problem solving process a number of different scaffolding approaches may be used.

The Centre for Interdisciplinary Science had previously developed a multi-stage, iterative problem solving strategy known as the *LEICESTER Strategy* (Raine & Symons, 2005) which has some similarities to the earlier Maastricht seven-step model (Maurer & Neuholm, 2012). The *LEICESTER strategy* is an iterative process designed to help students identify the nature of the problem, to audit their existing knowledge and to set group research targets. Due to the differences in the structure of the Chemistry and Interdisciplinary Science courses (e.g. the Interdisciplinary Science course adopted a short, fat structure whereas the chemistry PBL activities were integrated in a long, thin module), the *LEICESTER Strategy* was adapted to meet the needs of the chemistry course. A simplified version of the strategy was developed by focusing on some of the key aspects. The *SET Strategy* (Summarise-Existing Knowledge-Things to Research) was developed to encourage students to document their understanding of the nature of the problem, their summary of the relevant existing knowledge already held by the group and the group research targets (Williams et al., 2010). It was decided that every PBL problem would start with students making a written plan of the problem solving process based on this strategy. Students were instructed to update this document throughout the problem solving process.

The first challenge for students is to identify the nature of the problem. Students are encouraged to read the problem in advance of the PBL session so they come into the session with some thoughts on what is being asked of them and how they may set about solving the problem. During the PBL session, students are asked to construct a problem solving plan. The first part of this plan involves agreeing on a short (one or two sentence) summary of the problem. This is a useful peer-learning activity as it requires students to brainstorm ideas and then negotiate with fellow group members to decide on a suitable agreed summary.

Once this has been achieved, students must audit the relevant skills and knowledge of the group. If only some of the group members have particular skills or elements of knowledge, they are instructed to peer-teach the rest of the group before documenting this. A PBL problem will typically incorporate elements that students will not already be familiar with. In order to work on a solution to the problem, students must create a list of research outcomes that must be achieved by the group before the problem can be solved.

REDESIGNING TEACHING

A small group of staff and postgraduates were assigned to lead the teaching of the PBL component of the module. Training for those involved in PBL teaching was necessary due to the significant differences between this type of teaching (usually known as facilitation) and the teaching approaches that most physical science academics are familiar with (e.g. lectures, laboratory classes, tutorials, etc.). Facilitators support students through the problem solving process while allowing them to retain ownership of the problem solving strategy. The PBL philosophy is built around the principle of student-centred learning so it is essential that facilitators do not turn sessions into 'mini-lectures'.

The facilitator training course consisted of three sessions. The first session outlined the PBL method, the role of the facilitator and the specific details of how the approach integrated into the rest of the degree course. The second session was used to show facilitators how the Virtual Learning Environment (VLE) would be used to support the PBL process. The third session gave facilitators an experience of working in a group to solve a PBL problem. Getting facilitators to experience PBL as participants has proven to be a very valuable tool in allowing them to appreciate the difference between PBL facilitation and other forms of teaching. A written facilitator guide was prepared (which reinforced the points made in the training sessions, provided specific advice on how to support students working on individual problems and gave some suggestions of how to deal with potential challenges). Further support was provided in the form of weekly meetings of the facilitator group which allowed good practice to be shared and the discussion of any issues that had been experienced.

It was observed that postgraduates become better at PBL facilitation with experience (based on observation of their facilitation sessions and student feedback). New facilitators occasionally have difficulty knowing how and when to contribute to group discussions so it can be helpful to mentor them in the early stages.

REDESIGNING THE VIRTUAL LEARNING SPACE

Due to the group-focused nature of the PBL process, it became obvious at an early stage that much of the group work would be conducted outside of the timetabled sessions. In order to facilitate group collaboration outside of the timetabled sessions, a structure was created on the virtual learning environment that would allow students to collaborate remotely at any time. Students were given access to a group discussion board, a file exchange space and a group wiki.

The group wiki was the focal point of the entire PBL component of the module. Given the collaborative nature of wikis (i.e. easily editable, collaborative web pages (Parker & Chao, 2007)), it was decided that this format would be suitable for students to create and modify their solutions to the problems. Students were provided with training on how to use the virtual learning environment to collaborate with their group (in the form of a one hour 'Introduction to PBL' session and a set of written guidelines) and were given examples of good practice to refer to.

Using a wiki as the means of communicating the group solutions to the problems allowed regular feedback to be provided from facilitators (who could use the 'comments' feature to provide feedback on each wiki page) as well as from other group members (who could also use the 'Comments' features but could also edit and add to each other's work). Facilitators provided groups with regular feedback on both the quality of the academic work uploaded to the wiki and the effectiveness with which the wiki had been used (e.g. was it easy to navigate between different parts of the wiki? Had suitable hyperlinks been created between areas of related content?). Using a wiki also provides a mechanism by which facilitators can trace the level of contribution from different group members. It is not unusual for groups to appoint a single member to create the initial wiki structure and, in some circumstances, to upload the initial drafts of the problem solution to the wiki. Facilitators should expect there to be additional changes (e.g. edits made to the original content and additional content added) by other group members. At Leicester we do not use wiki contribution to help summative assessment of the individual contribution of each student (for the reasons described above) but we do provide formative feedback to students based on their level of contribution to the group problem solving process.

REDESIGNING ASSESSMENT OF GROUP WORK

Assessment of PBL activities can be challenging as it is often difficult to identify the exact level of contribution made by each individual member to the final group solution. Our implementation of PBL has been designed to provide an overall group mark (awarded by the facilitator) which is then scaled by a peer-review process and by the level of attendance of the individual in facilitation sessions.

Facilitators formatively assessed the group wikis by providing grades (see Table 1) for presentation and scientific content and by providing detailed feedback on a weekly cycle. Students were allowed to make further modifications to their solutions to the problems after receiving this initial formative feedback ahead of a final summative deadline. The assessment of the problems took a variety of forms including poster presentations, short oral presentations and written and drawn content which was presented on the group wiki.

Scaling of the group mark for individual members was achieved by collecting responses to a peer-review activity (hosted on the virtual learning environment). Students were asked to rate the performance of their fellow group members, these ratings were used to generate a scaling factor which helped fairly distribute the group mark to each individual group member. In the initial stages of our PBL implementation, facilitators also fed into review of individual contribution however this was changed from 2009 onwards as it was felt that this decision should be made by the students themselves.

EVALUATION METHODOLOGY

In order to measure the impact of the introduction of PBL, the following data was recorded:

- Student performance in a range of modules

- Student attitudes towards PBL in end of course questionnaires and focus group interviews
- Staff attitudes in focus group interviews

The data would be analysed and conclusions were to be used to help improve the approach for later years.

Table 1. Generic Marking Criteria for PBL Problems at Leicester.

Grade	Description
Content grade	
A	The answer contains all of the points listed and one or two extra related things.
B	The answer covers almost all of the points listed (only minor things missing) – any calculated values are close to the recommended values.
C	Most of the points listed appear in the answer, some missing content and/or some error in written content and/or calculations.
D	The answer has significant content missing. There are a number of errors throughout the answer.
E	The answer contains only a few of the important points from the list. There are significant sections missing.
Style grade	
A	The answer is very well presented and uses the appropriate format of delivery. Each section is virtually free from errors in grammar, spelling and punctuation and makes good use of *referenced and labelled* diagrams.
B	The answer is well structured and produced and uses the appropriate format of delivery. It is clearly written apart from relatively minor aspects which would not seriously affect the understanding of the reader. The answer makes good use of referenced and labelled diagrams to clarify key points.
C	Though reasonably well structured and produced the answer contains significant errors in grammar and spelling. Diagrams were provided, these were not always referenced and/or there was little attempt made to relate these to the answer. Some attempt was made to format the problem correctly.
D	The answer was difficult to follow for the reader. There were a number of errors – spelling, grammar and inconsistent use of fonts. Very few diagrams were provided and those that appeared may not have been appropriate. The answer was not formatted as was appropriate for the problem.
E	The answer was disorganised and disjointed and so badly produced that it would inevitably misguide the reader. There were lots of errors – spelling, grammar, lots of different fonts, and little or no evidence of teamwork. There was little or no effort to provide diagrams or examples to illustrate points in the answer. The answer was not formatted as was appropriate for the problem.

Note: Each problem has a list of essential points that must be included in the solution. These points are referred to in this table.

MODIFYING THE APPROACH TO MEET OUR NEEDS

This section will be used to present an overview of how the PBL implementation has been adapted over the last seven years to better accommodate the needs of both students and staff. One of the most significant early conclusions from the PBL implementation was that we needed to find an efficient way of running PBL so departmental resources were not overly stretched while still making this a productive learning experience for students.

ESTABLISHING A GROUP SIZE

A number of different facilitator: student ratios were attempted in the initial phase of the implementation. We found that a facilitator ratio of approximately 1:12 offered the most advantages in terms of student satisfaction. If a facilitator has to work with a greater number of students, there are lower levels of student satisfaction, if the number of students per facilitator is reduced, the cost of facilitating PBL sessions becomes too high to sustain.

SCHEDULING THE SESSIONS

There has been some discussion of the most suitable length of time that facilitation sessions should last. The Leicester Interdisciplinary Science course uses both two hour and one hour facilitation sessions. Our implementation of PBL has always been based on one hour facilitation sessions (partly due to timetable limitations and partly due to our intention to create a 'student-centred' focus where students would retain responsibility for arranging meetings between timetabled sessions). Initially, students were asked to attend two facilitation sessions per week but this was changed in 2009 to one facilitation session per week. Staff and student feedback indicated that students were being provided with ample time to discuss the problem and to receive feedback from their facilitator in the first of the weekly sessions. We have found that a single weekly session gives students enough time to discuss the background concepts (and the relationship of these concepts to the material they are covering in their lecture modules), to think about how to apply these concepts to solve the problem and then to reflect on the effectiveness of their decision and to set targets for work that needs to be done ahead of the next facilitation session.

In order to accommodate the required facilitator: student ratio, we were required to split the cohort across a number of separate timetabled facilitation slots. Between 2009 and 2013 each weekly facilitation session was split over four timetable slots. This made running PBL a relatively expensive teaching activity (compared to running a lecture course for example). In 2014 we were able to make use of additional teaching space in order to split the PBL facilitation over just two timetable slots per week. It can be difficult to retain a reasonable student: staff ratio, if there is a limitation of suitable teaching space so clearly much thought has to be given to this during the planning stage.

THE PHYSICAL LEARNING SPACE

Another concern when convening a PBL based course is ensuring that a suitable learning space is available for the facilitation sessions. PBL works best in a flat teaching space with furniture that can be configured to allow students to face each other (Raine & Symons, 2005). We have found that this configuration encourages all members of the groups to feel valued and to make contributions to the problem solving process. It can be useful to have the flexibility to change the configuration of the room as this allows the same space to be used for conference style poster and oral presentations that may form part of the assessment of PBL problems (Fig. 1). As many

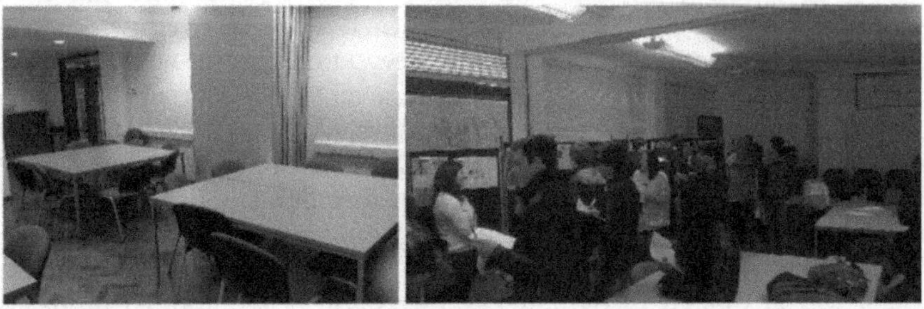

Fig. 1. An Example of a Flat Teaching Space used for PBL Facilitation at the University of Leicester. The Image on the Left Shows Group Work Layout and the Image on the Right Shows Poster Display Layout.

institutions have limited teaching spaces of this type, PBL teaching may need to be split over a number of timetable slots (and locations) in order to ensure that all students have access to a learning space which facilitates the problem solving process. We have also found that some thought needs to be given to the size of the space used for PBL facilitation. If the space is too small for the number of students in the session, the noise level can become disruptive.

One of the most significant changes in the way PBL runs at Leicester since its initial introduction in 2007 has been the role played by the introduction of 'Bring Your Own Device' (BYOD). When we first started using PBL most students did not own a laptop, a smart phone or a tablet device. Additionally some of the locations used for facilitation sessions between 2007 and 2010 did not have Wi-Fi access and none of the locations had any computers that were accessible to students. Since 2010 we have ensured that all rooms used for this activity have Wi-Fi coverage. This has allowed us to introduce a BYOD policy for all facilitation sessions. Throughout the first three years of our use of PBL feedback revealed the frustration students had in not being able to go online to research concepts during their group discussions. Since the advent of affordable smart phones and tablet technology, it has become very unusual for groups not have at least one member with an internet enabled device.

CREATING MORE AUTHENTIC PROBLEMS

As stated earlier, using or modifying existing PBL problems is a very good way of establishing the approach within a course. Once PBL has been established and the academic has decided what works in their own course, it may be time to consider developing a new set of PBL resources which is custom designed with the specific course in mind. We started developing our own PBL resources at the introduction phase between 2007 and 2009 alongside some 'imported' problems. In 2009 we removed the 'imported' problems which were replaced by new custom designed problems which aligned better with our intended learning outcomes. One of the driving forces behind this change was the lack of engagement from some students which was attributed to the unrealistic scenario. The imported problems were based on the premise that students were working as consultants to a Hollywood movie studio producing a sci-fi movie (based on a team of scientists who have been shrunk down to the atomic scale). The students had to communicate with a number of different people involved in the production of the film who had a range of scientific misconceptions. Student feedback comments revealed that they found this scenario '*trivial*' or '*insulting*' so the replacement problems were created with scenarios that had more realistic expectations of the types of situation that undergraduate chemists might be involved in (e.g. creating peer-learning resources on a particular topic for students on a different science degree course or solving an analytical problem at a small chemical company during an industrial placement).

Support was provided for the development of new resources from staff on Leicester's Interdisciplinary Science degree and a community of practice was established within the UK HE chemistry sector. Sharing ideas with colleagues at other institutions and possibly trialling resources on groups of students at other institutions can provide authors with very useful insights into the strengths and weaknesses of their new problems.

SUPPORTING THE PROBLEM SOLVING PROCESS

Over the last seven years major changes have been made to the way the facilitation sessions are structured. In the introduction stage of the project (2007–2009), students were given minimal formal guidance beyond the SET strategy and the feedback provided by their facilitator. Since 2009 a series of discussion questions have been introduced in each facilitation session. The aim of the discussion questions is to help focus the initial stages of the group discussion on the relevant areas of scientific content and to help students make connections between the problem and other parts of the course. The discussion questions have been written in a way that allows students to engage with the relevant areas of content without bypassing the student-centred nature of the problem solving process. In order to achieve this, the discussion questions are not usually directly related to the problems but will require students to consider some of the broader concepts and skills that they will need to apply in order to form a solution to the problem.

STUDENT INVOLVEMENT IN CURRICULUM DESIGN

One of the most recent changes to our approach has been the level of support we have given students in identifying their personal skills development.

One of the great advantages of the PBL approach is that it allows students to develop a range of transferable skills within the context of their academic studies. Before the introduction of PBL to the chemistry course at Leicester, opportunities for extended group projects and development of communication skills were limited during the early stages of the degree. The diverse nature of the assessment of the problems has allowed the integration of sophisticated workplace and transferable skills training within the chemistry curriculum. Students in later years of study tend to place a value on these skills, for example when applying for jobs or postgraduate study opportunities, but often struggle to relate these skills to what they did in earlier years of study. This has been addressed by briefing students about the transferable skills development opportunities afforded by this type of approach, and highlighting the fact that development of these kinds of skills at the early stages of a degree course can have advantages for students when presenting and evidencing their skills for job applications.

In order to support the development of an awareness of transferable skills development, a student–staff partnership has been created which is based around final year students working with first year students. Final year students have related the transferable skills developed in first year back to their own experiences of applying for jobs and completing their dissertation projects. This kind of peer mentorship can be more effective than further reinforcement of these skills from staff. Final year students created a series of online resources to help support skills development such as videos highlighting how the skills developed by the PBL part of the course will help students in a broader context and reflective questionnaires. Final year undergraduate students have also attended facilitation sessions as 'secondary' facilitators, there to help reinforce skills development during the problem solving process.

Since 2014 student partners have been used to help develop new PBL content. Students have been integral to the planning and development of new PBL resources. This process typically involves including final year undergraduate representatives in the planning and author teams. Throughout the 2014–2015 academic year, a team of final year undergraduate students worked on the development of a new PBL activity based on the interdisciplinary research area of food analysis. The student input was invaluable as it provided insight in to which parts of the problems would pose potential difficulties for students and also helped to develop an activity that students would find engaging. Another form of student partnership introduced in the 2014–2015 academic year was the enhancement of existing problems through the student-led development of pre and post session online quizzes to help focus student preparation and reflection. These quizzes provided students with an immediate feedback and helped students reflect on the subject related skills developed by working on the problems.

EVALUATION OF THE IMPACT OF PBL ON LEICESTER CHEMISTS

One of the primary concerns of academic staff ahead of the introduction of PBL to chemistry at Leicester was that the approach would be very time intensive for both staff and students (Heycox & Bolzan, 1991) and would possibly have an impact on student performance in other modules. The impact on staff workload has been discussed earlier in this chapter. The impact on student performance was measured by comparing student performance in a range of assessment types throughout the first year (including mid-term multiple choice assessments, tutorial work and end of module examinations) for year groups that had not done PBL (i.e. pre 2007) and those who had done PBL (i.e. post 2007) (Williams et al., 2010). The introduction of PBL had no measurable impact on student performance measured by any of the above methods. Students, on occasion, feedback that they find the workload associated with the PBL sessions heavy and that they believe this may have an impact on their performance in other assessments ("*PBL is a great idea, I just struggled with the overall workload this semester and other modules have suffered as a result*", University of Leicester undergraduate chemist, 2014) but there is little statistical evidence to support this.

One of the most significant early observations made during our initial implementation of PBL was the increase in retention rate that appeared to accompany the introduction of the approach (Williams et al., 2010). This increase in student retention (from 83% in 2006 to 89% in 2007 in spite of an increase student intake) was attributed to strong social links that students form when working in a small group of peers for the entire semester. This factor was of great importance to departmental staff when deciding to fully integrate PBL into the first year curriculum following the initial implementation.

REFLECTIONS

When reflecting on the changes made to the implementation PBL over the last seven years, it was felt that it may have been more productive to have started with a small scale implementation which was subsequently grown to meet the needs of the course. The department started with a large-scale implementation (which involved 20 hours of student contact time in a single term) which was soon found to be more time than students needed to achieve the intended learning outcomes. Starting with a small-scale implementation allows academics to identify the strengths and weaknesses of the approach without committing a large component of the course to the approach. Once academics have evaluated the small-scale implementation and decided how best to integrate PBL into their course, the implementation may be grown to encompass further intended learning outcomes.

Student reaction to the PBL approach is often very mixed as the approach is very different to what students are used to from the earlier studies. There have been some recent attempts to introduce to the PBL approach at further education level (Smith, 2012) but the number of such implementations remains small. Students sometimes have difficulty in accepting the fact that PBL problems do not usually have a single correct answer. Alternatively, students may also have difficulty working in a situation where much of the learning is directed by their own problem solving plan rather than according to an instructor directed plan ("*Objectives given in problems are vague, not enough information ... were provided*"). Both of these issues can be addressed by providing a transparent outline of the PBL process at the start of the activity and by reinforcing the mechanism of the PBL process through facilitator feedback. Taking responsibility for setting learning goals is a significant new responsibility for students so regular detailed feedback from the facilitator is essential.

Another one of the biggest student concerns about the PBL process is the impact that group members who fail to fully engage with the process will have on the mark awarded for the problem solution (*"PBL is good in theory, allowing us to explore content with context, but working as a group means we're dependent on everyone doing their fair portion"*). This issue can be addressed by creating a robust peer-review process which is reinforced by staff feedback on individual contribution levels throughout term. Students need to know that they will be penalised for not fully participating in the process and that if they suffer from having a non-participating group member, there are measures that can be taken to ensure their mark doesn't suffer.

Some students appreciate the benefits of the PBL approach ("*The PBL was useful for having to work in teams*", a student also commented that the PBL process "*mimics industry*") and clearly thrive when given the new responsibilities of planning a learning journey that will end with a solution to an open ended problem. In a 2014 questionnaire, first year chemistry students were asked whether they agreed with a number of statements following their experience of PBL. 63.2% of students agreed with the statement *Because PBL was done in teams it was a good way to meet people and make friends* and 60.5% of students agreed that *Doing PBL in teams was useful, I found it helpful to talk about some of the concepts with other students and PBL forced me to do this* (sample size = 38).

A student-partner generated questionnaire was used to measure which skills students thought they had used during the PBL activities and to measure their levels of confidence in these skills. The results are presented in Table 2.

Responses to the questionnaire clearly show that students appreciate the development of a wide range of transferable skills due to their experience of PBL. It appears that written communication (91.4% confident or very confident) and working in a team (97.2% confident or very confident) are the two skills which students most clearly think are developed. Application of scientific method (22.9% still not confident) and development of literature research skills (17.1% still not confident) are the two areas that had the highest number of 'not confident' responses. This will be addressed in the next academic year by ensuring that these activities are appropriately signposted in the problem documentation.

RECOMMENDATIONS FOR INTRODUCING PBL TO PHYSICAL SCIENCE COURSES IN HIGHER EDUCATION

This section provides a brief summary of the primary recommendations for introducing PBL in to a physical science course at higher education level.

Table 2. Responses to a Student-Partner Generated Skills Reflection Questionnaire for Year 1 Students. All Responses are Shown as Percentages. Sample size = 35.

	Literature Research	Scientific Knowledge	Problem Solving	Scientific Method	Written Communication	Record Keeping	Team Work	Time Management and Organisation
Practiced and very confident	14.3	17.1	25.7	14.3	45.7	37.1	48.6	40.0
Practiced and confident	68.6	71.4	60.0	60.0	45.7	51.4	48.6	48.6
Practiced but still not confident	17.1	11.4	14.3	22.9	8.6	2.9	2.9	11.4
Not used during PBL	0	0	0	2.9	0	5.7	0	0
No response	0	0	0	0	0	2.9	0	0

- *Use trained postgraduate facilitators*: adequately trained and supported postgraduate students can make effective facilitators. Interview potential facilitators to ensure they are comfortable with such an open ended teaching approach. Training sessions can be used to gauge how well potential facilitators cope with PBL problems. Use postgraduates who have some experience of the PBL approach where possible and retain trained facilitators for as long as possible.
- *Use or adapt existing PBL resources*: using existing resources can provide an excellent way to introduce the teaching approach in to a course. The problem can be adapted or replaced at a later stage if needed.
- *Try not to do too much at once*: When introducing PBL to a course it can be better to start with a small-scale implementation (as a form of 'pilot' introduction). Academics can reflect on this implementation and learn lessons that will help inform a larger-scale implementation.
- *PBL can be used effectively alongside other approaches*: PBL can be effectively integrated into a course which includes lectures, small group tutorials and laboratory classes.
- *Make use of student partners*: Undergraduate students can help develop new PBL resources or enhance existing resources. Feedback should be sought from students after every PBL activity.
- *Be realistic about what PBL can achieve*: PBL should not be thought of a solution to all problems. PBL will not necessarily improve exam performance (especially if the format of the exam does not reflect the PBL experience) and may not enhance student engagement in all cases.
- *Highlight professional skills development:* Ensure that problems address a range of professional skills and clearly signpost this skills development for the benefit of students.

REFERENCES

Andraos, J. (1999). A streamlined approach to solving simple and complex kinetic systems analytically. *Journal of Chemical Education*, *76*(11), 1578–1583. doi:10.1021/ed076p1578

Barrows, H. S. (1986). A taxonomy of problem based learning methods. *Medical Education*, *20*(6), 481–486. doi:10.1111/j.1365-2923.1986.tb01386.x

Belt, S. T., Enas, H. E., McCreedy, T., Overton, T. L., & Summerfield, S. (2002). A problem based learning approach to analytical and applied chemistry. *University Chemistry Education*, *6*, 65–78.

Blumenfeld, P. C., Soloway, C., Marx, W. M., Krajcik, J. S., Guzdial, M., & Palincsar, A. (1991). Motivating problem based learning: Sustaining the doing, supporting the learning. *Educational Psychologist*, *26*, 369–398. doi:10.1080/00461520.1991.9653139

Bredderman, T. (1983). Effects of activity-based elementary science on student outcomes: A quantitative synthesis. *Review of Educational Research*, *53*, 499–518. doi:10.3102/00346543053004499

Browitt, A., & Walker, L. (2007). *Retention and widening participation in the faculties of sciences and engineering*. University of Glasgow, Sutton Trust [Online]. Retrieved from http://www.suttontrust.com/wp-content/uploads/2007/10/1glasgow_science_and_eng.pdf

Duch, B. J., Groh, S. E., & Allen, D. E. (Eds.). (2001). *The power of problem based learning*. Sterling, VA: Stylus Publishing.

Glew, R. (2003). The problem with problem-based medical education. *Biochemistry and Molecular Biology Education*, *31*(1), 52–56.

Hendry, H., Harris, J., & Ryan, G. (2003). Group problems in problem-based learning. *Medical Teacher*, *25*(6), 609–616.

Heycox, K., & Bolzan, N. (1991). Applying problem-based learning in first year social work. In D. Boud & G. E. Feletti (Eds.), *The challenge of problem-based learning* (pp. 186–193). London: Kogan Page.

Maurer, H., & Neuholm, C. (2012). *Problems everywhere? Strengths and challenges of a problem-Based learning approach in European studies*. Retrieved from http://www.mceg-maastricht.eu/pdf/MCEG_part%20C/PBL_link2_%20C/PBL%20implementation%20challenges.pdf

Newman, M. J. (2005). Problem based learning: An introduction and overview of the key features of the approach. *Journal of Veterinary Medical Education*, *31*(2), 12–20.

Overton, T. L., & Randles, C. A. (2015). Beyond problem-based learning: using dynamic PBL in chemistry. *Chemistry Education Research and Practice*, Advance Article. doi:10.1039/C4RP00248B

Parker, K. R., & Chao, J. T. (2007). Wiki as a teaching tool. *Interdisciplinary Journal of Knowledge and Learning Objects*, *3*, 57–72.

Raine, D. J., & Symons, S. L. (2005). *PossiBiLities: A practice guide to problem-based learning in physics and astronomy*. York: The Higher Education Academy – Physical Sciences Centre.

Savery, J. R., & Duffy, T. M. (1995). Problem based learning: An instructional model and its constructivist framework. *Educational Technology*, *35*(5), 31–38.

Smith, C. J. (2012). Improving the school-to-university transition: Using a problem-based approach to teach practical skills whilst simultaneously developing students' independent study skills. *Chemistry Education Research and Practice*, *13*, 490–499, doi:10.1039/C2RP20096A

Williams, D. P., Woodward, J. R., Symons, S. L., & Davies, D. L. (2010). A tiny adventure: The introduction of problem based learning in an undergraduate chemistry course. *Chemistry Education Research and Practice*, *11*, 33–42. doi:10.1039/C001045F

Wood, J. C., & Mack, L. G. (2001). Problem-based learning and interdisciplinary instruction. *Proceedings of the 2001 American Society for Engineering Education Annual Conference & Exposition*, Session 1648.

Woods, D. R. (1996). *Problem-based learning: Helping your students gain the most from C/PBL*. Hamilton, Ontario: McMaster University.

AN INTENTIONAL AND FACILITATIVE STUDENT-CENTERED APPROACH TO ENHANCE INQUIRY-BASED LEARNING THROUGH AUTHENTIC, HYPOTHESIS-DRIVEN RESEARCH

Amelia J. Ahern-Rindell

ABSTRACT

My interest in inquiry-based learning spans my career teaching science and non-science majors about cell/molecular biology, genetics, and evolution. My Genetics Lab course allows students to perform hypothesis-driven research in small groups using my research program as the basis for their authentic, discovery-based experiences. This course has undergone multiple revisions and is now suitable for the cognitive developmental level of undergraduates; it challenges them but avoids frustrating them. I will share how to navigate the trade-off of reducing content

Inquiry-Based Learning for Science, Technology, Engineering, and Math (STEM) Programs: A Conceptual and Practical Resource for Educators
Innovations in Higher Education Teaching and Learning, Volume 4, 113–134

ISSN: 2055-3641/doi:10.1108/S2055-364120150000004008

coverage in favor of emphasizing the scientific process and the actual doing of science. I will describe how to help students understand the difference between information and knowledge, how to better comprehend the primary literature, and how to build on this body of work to advance scientific discovery and understanding. I will illustrate how to scaffold and improve student's existing skills and abilities such as critical thinking, oral and written communication, data interpretation and reporting, and how to develop and expand on their creativity in posing questions that become the basis of their hypothesis-driven work in the laboratory. Finally, I will explain why I choose to use my research program as a means of guided discovery rather than an open-ended format.

As a teacher-scholar for more than 20 years, it is no surprise that my research endeavors help inform my teaching and allow me to stay current in the classroom in my field of expertise. However, one unexpected and fortuitous happenstance that also occurs is how my research program reciprocally benefits by the interactions I have with the students in my courses. The give and take of teaching and learning often blurs the designated teacher/student roles making us instead, companions in the act of acquiring knowledge. For my generation, "the sage on the stage" or the teacher-centered model describes how the teacher was the only active body in the classroom (King, 1993). The students sat passively at their desks with all their eyes fixed on the professor at the front of the classroom who seemed to perform for the students as in a play. The professor would share their knowledge in long soliloquies and expected the students, through osmosis or some magical phenomenon to not only retain this dispensed wisdom, but with no assistance what so ever be able to apply it in an appropriate manner.

Fortunately for the students of today, the teacher-centered classroom is dying, albeit a slow death, and instead, teachers are becoming facilitators of, and companions in their student's learning, that is, "the guide on the side" (King, 1993). With the vast amount of information that is now available at a touch to the keypad, the professoriate has come to realize, and put into practice, various ways to help students critically analyze the glut of information at their fingertips. Classrooms are becoming student-centered and utilizing different learning pedagogies like problem-based learning (Hmelo-Silver, 2004). Faculty must do a better job of modeling for their students how to incorporate that which is credible into their knowledge base and demonstrate how to use this newfound wisdom to answer pertinent questions and solve challenging problems. That is not to say that the lecture format does not have its benefits. To learn more critical thinking skills and how to problem solve, my experiences and those of many others suggest that students definitely need to see this behavior demonstrated for them and to have opportunities to apply knowledge in addition to being exposed to it in a lecture, in a text, or through some other more modern format (Renkl, Atkinson, Maier, & Staley, 2002).

EVOLUTION OF AN IDEA

My interest in inquiry-based learning spans my career as an undergraduate professor teaching science and non-science majors in the sub-disciplines of cell/molecular biology, genetics, and evolution. In addition to my lab-oriented research, I also engage in the scholarship of teaching. My experiences in the classroom inform curricular and pedagogical changes that I continually make and then share with my peer educators. Through course evaluations and self-assessment, I continue to redesign my classes to include activities that promote student learning rather than rote memorization of facts. This practice began early in my teaching career when I received funding from the National Science Foundation's *Course, Curriculum, and Laboratory Improvement Program* in 1994 (National Science Foundation, 1994), and four years later when I changed institutions (National Science Foundation, 1998). What I proposed was the implementation of newly designed laboratory courses that utilized an investigative and collaborative student-centered approach that sought to improve conceptual learning by engaging students in doing real science rather than "cookbook" experiments. I made the argument that these canned activities do not adequately illustrate the scientific process and provide little challenge or novelty, which is what is so exciting about doing science, and the major reason for becoming a scientist in the first place.

My proposed course model would allow students to perform hypothesis-driven research in small groups using my own research program as the basis for their authentic, discovery-based experiences. This pedagogical strategy has many benefits for all the stakeholders, including students, faculty, institutions, and society. However, it is only after years of trial and error that I developed a framework that I think can be shared with others to enhance and effectively nurture inquiry-based teaching and learning.

One of my goals as a teacher has been to improve the educational experience in the classroom by deliberately moving away from the teacher-centered model towards a more intentional and facilitative approach. The ideal is to model the appropriate behavior for my students to emulate, and then provide opportunities for them to practice the skills necessary to become independent, critical thinkers and lifetime learners.

STRUCTURING AN INQUIRY-DRIVEN COURSE

It is my intention to provide the reader with a practical example of a science-oriented, inquiry-based learning experience modeled from my current Genetics Lab course. This course has undergone multiple revisions based on my cumulative teaching experience, student feedback from course evaluations, insights from undergraduate teaching assistants, and input from student research collaborators who began independent projects under my mentorship after taking previous permutations of this class. These different perspectives helped me improve the course's framework, which is now more suitable for the cognitive developmental level of undergraduates. It challenges them but avoids frustrating them, which would be counter-productive and not conducive in maintaining a supportive learning environment.

I will proceed by sharing an explanation of how to navigate the trade-off of reducing content coverage in favor of emphasizing the scientific process and the actual doing of science. Deciding what content should be explicitly shared should in part be determined through a student participatory activity that uncovers, and then proceeds to eliminate, student misconceptions about the subject area, in this case, how authentic science is conducted in general, and in the genetics discipline in particular. I will describe how I help students understand the difference between information and knowledge, how to better comprehend the primary literature, and how to build on this body of work to advance scientific discovery and understanding about how the world works. I will illustrate how I scaffold and improve upon student's existing tool kit of skills and abilities such as critical thinking, oral and written communication, data interpretation and reporting, and how to develop and expand on their creativity in posing questions that become the basis of their hypothesis-driven work in the laboratory. Finally, I will explain why I choose to use guided discovery by utilizing my research program as the foundation for our work together rather than an open-ended format. This not only illustrates the collaborative nature of research in the sciences by having them working with me on what I am passionate about, but it also models for them how to create productive partnerships through my personal experiences. Since I am intentional when assisting them with forming their student groups, and they know ahead of time I will be asking them to critique their own and their group members' performances, they are very aware of the need to maximize their interactions and reach a balance in supporting their collaborative efforts. I, in turn, by helping them form their groups by using a short survey, which probes how each student views their own work style, can influence and help manage the group dynamics by attempting to minimize the potential for social interpersonal conflicts.

QUALITY VERSUS QUANTITY

The natural sciences are a group of fact-laden disciplines that are content-rich and jargon specific, especially biology with all of its sub-disciplines. It is not practical to believe that a semester-long undergraduate lab course in Genetics can even scratch the surface of this subject's foundational concepts let alone provide sufficient content to flesh out the major gaps in our knowledge. From an inquiry-driven perspective, one cannot expect to delve into the consequential questions whose answers will help advance a biology sub-discipline if one is preoccupied with exhaustively "covering the content." It is acceptable to sacrifice quantity – an overabundance of facts, if the end result is quality – improved comprehension, application, and synthesis of a few fundamental and representative concepts. Less content means time can also be spent on the acquisition of skills that will prepare students for more advanced educational training, immediate employment, or life-time learning after receipt of their baccalaureate degrees. Possessing knowledge that you cannot apply to solve problems, or use that will lead to future discovery and the creation of new facts that improves understanding, is just a collection of useless truths.

The *Vision and Change in Undergraduate Biology Education Report* that was sponsored by the National Science Foundation, the National Institutes of Health, and the American Association for the Advancement of Science speaks directly to this issue (Vision & Change in Undergraduate Biology Education: A Call to Action, 2011). They say rather concisely and quite eloquently in this report, "... we do not attempt to 'cover' all there is to know.

Indeed, given the explosion of new knowledge in the biological sciences, attempting to 'cover it all' can be counterproductive and turn away even the most talented and interested students" (Vision & Change in Undergraduate Biology Education: A Call to Action, 2011, p. 13). My Genetics Lab course, which has been under development for the past 20 years, epitomizes the recommendations recently articulated in this Vision and Change Initiative. As a faculty member, it is reassuring to know that there is a respected support for organizing your course on a few chosen themes that succinctly capture the foundational breadth of your discipline, which then permits sufficient time for their in-depth and thorough coverage.

Articulating for your students a few major concepts that you can expand upon and tie together in a synergistic manner provides the skeletal framework from which a student can add to as they take other related courses and/or venture out on their own into the primary literature. By relating classroom activities back to the course learning objectives (Box 1) and intentionally highlighting contextual connections for our students, encourages and prepares them to make a concerted effort to look for these associations on their own.

In retrospect, my naïve hope that students would discover for themselves these interdependent links, without me ever mentioning this expectation, denies the student the opportunity of ever achieving this ability. This skill, which I would describe as being "insightful," does not happen on its own. It too must be recognized as an acquired ability that requires a demonstration of the modeled behavior first and opportunities, initially some of which are more deliberate, for ongoing practice. Acquiring the skills that lead to deep learning (Kuh, 2008) such as possessing insight, is essential to enriching a student's future learning potential and makes the course content at hand more cohesive and integrated, and thus more meaningful to the student.

When introducing the major concepts to be covered in a course, it is critical that student's existing knowledge of the concepts be determined to ensure they have an accurate base to build upon. If misconceptions exist, they must be dealt with and eliminated to avoid confusion and conflict that prevents the student from moving beyond their false perceptions. The teacher must identify and help correct the fallacies before she/he can hope to layer on additional knowledge. If the foundation is faulty, what is built next will not be supported and will likely not be retained or stored away in a usable form. This prevents additional growth and the opportunity for a student to achieve deeper understanding with a greater appreciation for the

Box 1. Genetics Lecture Course Learning Objectives and Learning Activities.

The student will demonstrate:

1. An understanding of genetic terminology and principles by applying them in appropriate situations. (*All class activities, assignments, and exams*)
2. An understanding of DNA's molecular structure and how it relates to the concept of the gene, the chromosome, and an organism's genome. (*Case studies, problem sets*)
3. An understanding of DNA's functionality including how it stores information, replicates and repairs itself, and the mechanisms by which it is regulated, expressed, and evolves. (*Case studies, problem sets*)
4. An understanding of the relationship between DNA, RNA, and protein and how these biological molecules contribute to an organisms' physical and psychological traits. (*Case studies, problem sets, pedigree analysis*)
5. An understanding of probability and how it is used to determine trait expression and assess risk for the occurrence of genetic disorders. (*Case studies, problem sets, Chi Square analysis*)
6. The ability to construct and interpret pedigrees and differentiate between different modes of inheritance. (*Case studies, problem sets, pedigree analysis*)
7. An understanding of the historical basis of the genetics sub-discipline (Mendel's contributions for example) and the significance of "classic" genetic experiments by showing its relevancy to the genetics field today. (*Problem sets, pedigree analysis, Chi Square analysis*)
8. An understanding of the role genetics plays in contributing to the unity and diversity of life. (*Primary literature*)
9. An understanding of the importance of bio-diversity by describing the harmful consequences of a lack of diversity when placed in the context of genetic variation. (*Case studies, primary literature*)
10. An understanding for the scientific process of inquiry and its use to solve problems relevant to genetics. (*Case studies, primary literature, Disorder ppt presentations*)
11. An understanding of how genetics informs other sub-disciplines of biology including development, physiology and anatomy, microbiology, animal behavior, botany, zoology, and most definitely, evolution. (*Case studies and problem sets*)
12. An understanding of how genetics is informed by mathematics, computer science, and the physical sciences. (*Problem sets, DNA sequencing analysis, bioinformatics exercises*)

nuanced complexity of a concept or its interconnectedness to other major concepts in the course.

The Genetics lecture course (85 minutes twice a week, ˜40 students) and the Genetics lab course (3 hours once a week/section of ˜20 students maximum) are co-requisites of one another. This allows coordination of the content and activities that provides a more synergistic experience. For example, at the end of the semester we use the lecture classroom for our Research Poster Symposium that allows the two lab sections to come together to interact in a venue that resembles what one experiences while attending a disciplinary conference. The melding of the lecture and lab courses results in a more cohesive and engaging format that allows these two genetic courses to re-enforce each other and permits the lab to be more process-focused that better emulates authentic scientific research.

AN AUTHENTIC RESEARCH EXPERIENCE

The Genetics Lab course is a microcosm of the scientific process. During the semester students partake in all facets of the scientific method including formulating a hypothesis and designing an experimental plan that is presented as a research proposal, learning about and utilizing molecular techniques to gather data, performing data analysis and interpretation, drawing conclusions from the data, redesigning experiments and gathering additional repeatable data, and then disseminating their research findings with the creation of a poster and its presentation during a symposium. Up until this course, our students are exposed to the various steps of the scientific process conceptually or through disjointed hands-on experiences in our general biology introductory 4-course series. However, what makes the Genetics Lab course different, which is substantially more challenging for students and inherently more valuable, is combining the conceptual framework and all the actual doing together in one course as they investigate an authentic research problem. Not only does this bring the scientific process to life, but it also becomes a way of knowing for the students and they begin to appreciate the reality of becoming a scientist and what doing research is really like.

Engaging undergraduates in a research experience of some kind is considered one of the high-impact practices associated with deeper learning, which I referred to previously (Kuh, 2008). Deeper learning is a developmental process wherein students learn the skills of critical thinking that go beyond comprehension and conceptual understanding to more complex abilities such as application, integration, and synthesis. The Council on Undergraduate Research (CUR), defines undergraduate research as "an inquiry or investigation conducted by an undergraduate student that makes an original intellectual or creative contribution to the discipline" (Council on Undergraduate Research [CUR], 2011). Assessments of undergraduate research describe the many benefits to students and how they learn about the scholarship of their discipline (Laursen, Hunter, Thiry, & Melton, 2010; Lopatto, 2003, 2004). Research experiences can provide academic challenge and an enriching educational environment if they are done right. Utilizing an authentic research scenario allows students to see what research is really like rather than performing some canned cookbook facsimile.

One thing the Genetics Lab course allows for is the opportunity for students to experience how to troubleshoot their experimental design. A critical aspect of performing scientific research is the iterative nature of the process. This is not just in the form of repeating an experiment to gather sufficient data for statistical validation, but what must occur early on to perfect the experimental design and ensure the hypothesis being tested is in actuality what the experiment is evaluating. Troubleshooting a protocol because you failed to obtain any results, did not obtain what was predicted, or the results just seem counter intuitive is truly at the heart of doing experimental research. This is the challenging dimension to conducting good science, and not allowing students to experience this part of the scientific process, deprives them of appreciating the individuality and creativity that is a key element of experimental design. This troubleshooting aspect of research directly leads to obtaining credible results versus ambiguous data from which no concrete conclusions can be drawn. This is where students either become hooked and envision graduate school as a next step, or realize a mismatch regarding their initial interest in what they perceive science to be versus the reality check of what day-to-day science is really like. All of this can best be appreciated through the aforementioned active participation that takes place during an authentic research experience.

On day one with the recently up-dated, instructor written lab manual in hand, students are introduced to my research program. The preface of the lab manual outlines what is ahead and specifies my high expectations of the students with a clearly articulated message of this being a class unlike any other that the students might have been exposed to previously. By fully informing the students of what will happen in the course, and that they will be asked to go beyond their comfort zone, inspires them to rise to the occasion. It compels them to be more open-minded and comfortable with the notion that none of us knows what the outcome of their research will be. The students find it very exciting and intriguing that even I don't know the answer to their research question. It pushes them to reach past their usual level of achievement and results with them taking ownership of their group projects. Although the lack of a major finding often leads to a bit of disappointment, the experience none-the-less has a strong impact on the students such that they comment in their course evaluations using words such as transformative, empowering, or enlightening.

COURSE FORMAT AND IMPLEMENTATION

As with any lecture or lab designated course, using a very detailed syllabus that acts like a contract, informs the students of what will happen when, and what is expected of them. The syllabus and lab manual each have a specific role to play regarding the information they provide and are good examples of how the means of delivering the message can often contribute to its understanding. For example, the learning objectives for the course should be clearly articulated in the syllabus with an explanation of how the students will be exposed to the knowledge they will be expected to acquire. The learning activities, by which they will practice applying the provided facts, in order that they might incorporate it into their knowledge base for current and future use, should also be included (Box 2).

Additional organizational issues that I worked on over the years to improve upon the structure and the formatting of the course are described below in the section "Course Logistics." Then the section "Communicating Science" pertains to written and oral means by which scientists share knowledge with one another for peer-review purposes. The last section, "Closing the Loop," covers how I provide prompt feedback to the students on assignments, and for assessing the course itself to continue to improve it.

Course Logistics

Good communication is one of the most important aspects of any course but when it comes to safety concerns it is much more critical for lab courses. I try to use as many modes of communication as possible, choosing the most suitable based on the type of information to be conveyed versus the

Box 2. Genetics Lab Course Learning Objectives and Learning Activities.

The student will demonstrate:

1. An understanding of genetics principles and terminology and apply them in appropriate situations. (*All lab activities, assignments, quizzes, and exams*)
2. An understanding of how to engage in the scientific process which includes formulating hypotheses, designing experiments, analyzing and interpreting data, and drawing reasonable conclusions. (*Research proposal and lab activities*)
3. Critical thinking and problem solving abilities. (*Research proposal, poster presentation, primer design, and sequence analysis*)
4. Effective oral and written communication in a scientific context. (*All lab activities, lab notebook, all assignments, quizzes, exams, primary literature, and poster presentation*)
5. Quantitative reasoning and the ability to apply statistical analyses. (*DNA concentration determination, restriction digest mapping, data analysis and exams*)
6. Information literacy through the ability to analyze the scientific literature and to write in a scientific manner. (*Research proposal, poster presentation, primary literature, and journal article analysis*)
7. Competent use of equipment and experimental techniques to gather data. (*All lab activities*)
8. An understanding of the responsibilities inherent in perpetuating and applying knowledge. (*All lab activities and research proposal*)
9. Ethical reasoning and good judgment in assessing the social and economic forces that influence scientific research. (*All lab activities and research proposal*)
10. An understanding of what scientific integrity is through appropriate behavior in a research environment. (*All laboratory activities*)
11. That he/she can work collaboratively in groups by sharing responsibilities equitably and that she/he can utilize one another's strengths to problem solve. (*All lab activities, research proposal, and poster presentation*)
12. That he/she understands the benefits and consequences of shared experiences by engaging in safe conduct in a research laboratory environment. (*All lab activities*)

urgency of the communiqué. I use a convenient *email system* for all my classes that my institution provides and supports. An *online course-specific tool* for posting materials and allowing students to upload and submit assignments is also available. The *syllabus* is posted on the course site the week prior to the beginning of the semester to reinforce the importance of the document and to provide time for students to read and digest prior to our first meeting together. The first day of lab, the students are reminded to read the syllabus in its entirety and to ask questions as needed. During our second class, students are expected to sign a form attesting to the fact that they read the syllabus and will abide by its guidelines and those presented in the lab manual I have written for the course.

The *lab manual* is revised every summer prior to the course being taught in the fall each academic year. Since the course content is based on my research program, which is naturally dynamic, timely updates are essential to provide the authentic research experience (see below). In addition to protocol/technical information, the lab manual also serves as a resource providing appendices of equipment and supplies that will be used, online bioinformatics tools and free-source websites that are relevant, pertinent primary references that are applicable to the research focus, weekly questions and problem sets to reinforce and provide opportunities for practice and skill building, and safety information.

Safety issues are addressed the first day of class and every meeting thereafter. Students are provided a checklist that they use as we tour the laboratory and I point out the locations of the safety equipment and demonstrate proper usage. The students are required to sign the form to stress the importance of the information. At the beginning of each class the pertinent safety issues are re-emphasized during the pre-lab introduction.

The pre-lab introduction is used to make sure all students know what we will be doing during the class. I use a short *power point* (that is uploaded to the course site in advance) that we go over first as a review of the relevant information including: safety issues, demonstration of equipment use, a diagrammed explanation of what the reagents and supplies are on their lab benches, and what we will accomplish. The student is expected to be prepared with the written entries in their *lab notebooks* consisting of concise flow charts of that week's procedures. The protocols and techniques that will be used are thoroughly outlined and illustrated in their lab manuals and do not need to be rewritten. Instead, students are asked to create a section in their lab notebooks dedicated to these flowcharts that help provide the big picture overview so students can readily comprehend what they are trying to accomplish and coordinate their efforts with their group members.

Lab notebooks may be a new entity for many of the students and require a detailed explanation. In spite of including a thorough discussion of how to keep a lab notebook in the course's lab manual, it doesn't usually sink in until they read my feedback after I return the collected lab notebooks partway through the semester. Keeping a good lab notebook is very important to the research endeavor and therefore is a topic students are asked about on the course midterm.

An *exam* is given halfway through the semester just prior to the students beginning their group research projects. The exam covers everything we have gone over in the first seven weeks of the course. This is a problem-based exam that allows students to apply the knowledge that they have acquired and incorporated into their working memory. It provides a good means of determining if any student needs additional help with the content or skills prior to them becoming more independent in the laboratory as they work on their group projects.

Communicating Science

Helping students navigate the *primary literature* from a comprehension standpoint and as a resource is critical if they are ever going to become comfortable with, and truly cognizant of, the scientific process and how it works. Appreciating the differences between a primary versus a secondary reference is the first major hurdle. Exposing students to examples of each is critical to them eventually being able to distinguish one from the other. In the Genetics Lecture course students are asked to read the Tuesday edition of the New York Times Science and Health sections in search of articles that are relevant to the field of genetics. They choose one of the articles they find during the semester and write a short analysis paper on it addressing the issue of how accurate the secondary source is in reporting genetic discoveries or advances and whether there is any indication of bias. This assignment allows for a means of comparison to the primary articles we then look at in the lab course.

In the Genetics Lab course the students read a couple of primary references pertaining to my research program during the first few weeks of the semester. They answer a series of questions about each of these papers to determine their level of comprehension and their ability to critically analyze how well the authors' conclusions are supported by the data. We work through the first paper together in class after the students have made a good faith effort to read and analyze the paper on their own. For the second paper, they are asked to work on the analysis alone and submit it as a graded assignment. I return these graded analyses with my detailed comments and we discuss them as a class for additional reinforcement. After these guided experiences the students are tasked with including at least five primary references into their group research proposals in which they must demonstrate their comprehension of the journal article and how it contributes significantly in supporting some foundational aspect of their research proposal.

For a majority of the students, writing a *research proposal* is a new experience. Each group collaborates in writing a complete first draft that includes an informative title, an abstract, an introduction with their hypothesis, purpose, and significance, the materials and methods, a section on predicted results and an interpretation of how these possible results will impact revision of their experimental design, and a reference section of the in-text citations. After reviewing and incorporating my substantial written comments/suggestions, and having the benefit of a face-to-face feedback exchange, the revised final version of the proposal is much improved and indicates a level of sophistication that was not present previously.

This same level of improvement is also noted with the *lab notebooks* that are collected for a second time at the end of the course. The areas that were initially very weak such as a lack of detailed experimental observations, insufficient data analysis, and marginal conclusions that were devoid of reflection and insight now convey a more careful and deliberate effort on the part of the students that make their entries more thorough and perceptive. Both of these examples, where students have the opportunity to utilize feedback to improve upon their initial works, speaks to the benefit and value of the iterative process for learning, and in these specific cases, how to write in the science discipline.

This enhanced writing ability is further demonstrated in the posters that the students create at the end of the semester in preparation for the poster symposium when they present their research findings to their peers and invited guests. An integral step to the students preparing well thought out posters is the experience they have during the first class period when in small groups they evaluate posters from the previous year's lab sections. The students are asked to answer a series of questions that walk them through the different sections of their assigned poster. This serves as an example of a primary reference and as a model demonstrating good as well as bad poster elements that address issues of content, format, and aesthetics. This mock peer-review exercise accomplishes four things. First, it begins the process of their familiarization with the *background knowledge* they will need to understand the problem that they are being asked to help solve. Second, it allows them to see the *finished product* that they are expected to produce making them well aware of the work that is ahead of them. Third, it prepares them to *think critically* setting the stage for the primary paper analyses that they will work on in the coming weeks. Finally, it plants the seed and challenges them to think about what they will do differently to *design a poster* that will better convey the problem to be solved and what they contributed to its resolution.

Closing the Loop

In keeping with the concept of critiquing another's work as well as our own, this last section on course format and implementation describes the ways in which I use various *evaluation tools* to close the loop on the activities the students are engaged in during the semester. The first and most important from a standpoint of providing prompt feedback for students is the use of assignment evaluation forms that are created to be specific for each graded assignment, but have common components that are universally assessed in the course regardless of the assignment (Box 3).

These evaluation forms are included in the syllabus so there is no question in the student's mind on what they will be graded, why, and when. In combination with the grading rubric for the course, the student is well informed about expectations and the consequences when they are not met (Box 4). This makes it very clear that when they do not meet the required quality standards, and/or do not act in accordance with the specified assignment guidelines, that they will earn a lower grade that reflects these deficiencies.

This prompt feedback is critical if a student is going to change their behavior and be able to implement the suggested changes on the future assignments, which is indicative of what we define as "learning." Reading comments and suggestions written directly on their assignment provides a very specific feedback on that particular task, which is valuable but not sufficient to significantly change future behavior. This type of feedback is only a small part of what students need and deserve. The true enlightenment and empowerment that students can experience is through a bigger picture perspective that a grading tool like an evaluation form can provide. It puts

Box 3. Genetics Lab Course Research Proposal Evaluation Form.

Research Proposal (draft and final)
This is the form that will be used to grade the draft and final versions of the research proposal that is due at the beginning of class **WEEKS 5 and 9, respectively**. The form indicates the areas you will be evaluated on and the points possible for each area (first number is draft and second is the final). What should be contained in each area of the proposal is outlined in the syllabus. Please consider these criteria as you write your research proposal.

Content:
Title (1, 2)
Abstract (2, 4)
Introduction and Background (4, 8)
Purpose and Rationale (3, 7)
Materials and Reagents (2, 4)
Proposed Methods (and Statistics) (4, 8)
Possible Outcomes and Interpretations (3, 7)
Literature Cited (2, 3)

Coherence:
Organization (1, 3)
Paragraphing (1, 3)
Transitions (1, 3)
Word Choice (1, 3)

Mechanics:
Sentence Structure (1, 3)
Grammar (1, 3)
Spelling (1, 3)
Punctuation (1, 3)
Followed Guidelines (1, 3)

Content (21, 43)____
Coherence (4, 12)_____
Mechanics (5, 15)_____
Total (30, 70)_____

Additional Comments:

Box 4. Genetics Lecture and Lab Courses Grading Rubric.

Grading Rubric for oral and written work:

A: Student conveys an **excellent** understanding of the concepts and terminology regarding the topic. Ideas are expressed **clearly** and **succinctly** in terms that are appropriate for the target audience. The content is scientifically **accurate**, its presentation is **original, perceptive,** and **well organized**, and the spelling, punctuation, and grammar are **flawless**. Student has incorporated cited references **appropriately**.

B: Student conveys a **good** understanding of the concepts and terminology regarding the topic. Ideas are expressed **clearly for the most part**, but some terms or concepts are **not explained thoroughly**. The content is scientifically accurate, but the presentation **lacks** the perceptiveness or originality of an "A" paper. Contains only **minor** errors in spelling, punctuation, or grammar. Student has incorporated cited references appropriately **for the most part**.

C: Student conveys a **satisfactory** understanding of the concepts and terminology regarding the topic. Ideas are expressed coherently, but **not organized** in a clear and concise manner. Important terms or concepts are **not explained sufficiently**. Contains **minor** scientific **inaccuracies** or **lacks** significant pieces of information. Contains only **minor** errors in spelling, punctuation, or grammar. Student has incorporated **some** of the cited references **inappropriately**.

D: Student conveys a **weak** understanding of the concepts and terminology regarding the topic. Ideas are expressed **intelligibly**, but **not organized** in a coherent manner. Content represents **minimal** effort by the student. Important terms or concepts are **not explained**. Contains **significant** scientific inaccuracies or lacks important pieces of information. Contains **frequent** errors in spelling, punctuation, or grammar. Student has incorporated **some** of the cited references **inappropriately or not** used a sufficient number of references.

E: Student conveys **little or no** understanding of the concepts and terminology regarding the topic. Presentation contains **incoherent** statements or **lacks** meaningful organization. Important terms or concepts are **omitted**. Contains **significant** scientific **inaccuracies** or **lacks** essential pieces of information. **Frequent** errors in spelling, punctuation, or grammar make the paper difficult to understand. Some information is presented **poorly** and/or is **irrelevant** to the topic. Contains **plagiarized** text. Student has incorporated **many** of the cited references **inappropriately or not** used a sufficient number and/or type of references.

the assignment-specific "errors" into an overall context explaining how they detract from the quality of their work. The evaluation form provides the why in conjunction with the what, which is process-oriented rather than focused on individual mistakes.

This impacts the grading process overall. It allows the faculty to be more consistent with their measure of quality from each student's assignment to the next, and it continually reminds faculty to be deliberate and intentional in using language that brings the message back to the process that the assignment was designed to address. The evaluation form also keeps faculty focused on the particular learning objective(s) each assignment is intended for so feedback can be more purposeful, precise, and explicit.

Opportunities for the students to provide feedback about the course are also essential for assessment purposes allowing one to "close the loop" by making changes to the course for the next time that it is taught. At my institution, the student's feedback also comes in the form of online, evaluation surveys to which I can add questions of my own that can be made specific for each course. There are always five standardized questions that attempt to examine the course and the faculty's role. They include whether the course was a valuable learning experience, if the instructor was an effective communicator and conveyed content well, whether a helpful learning environment was maintained, and if the instructor was available outside of class. To increase response rate, which helps to prevent skewed data from only the extreme perspectives, I make time available in our last class session for the students to work on these online evaluations. Some of the questions I ask the students pertain to critical thinking skills involving analysis, synthesis and evaluation. Those students that respond do so in a positive manner indicating that the course helped them with these skills and increased their comprehension of the discipline-specific content that was covered (Box 5).

The last form of critique that the students are asked to participate in pertains to the functionality of their groups. They are asked to reflect on their groups' dynamics overall and on each individual's contributions or lack thereof. These written evaluations are usually very positive about the group members' interactions and how they supported one another. They recognize that they learned from one another and the importance of good communication and equal sharing of the workload. They seem to understand that everyone has strengths and weaknesses and they tried to capitalize on this to improve their efficiency and productivity. On occasion there will be some groups that recognize when they did not function as cohesively

Box 5. Genetics Lab Course Student Evaluation Responses.

Do you think you improved your synthesis skills such as taking what you learned and using it to produce a new product such as a paper, a research proposal, or a presentation that requires original, creative thinking?

1. Yes. Making experiments into posters to present.
2. Yes, not only plastering immediate results onto a poster, but go a bit further to get all the info, out of those results as possible.
3. –
4. Yes, having the model of what to do and then actually doing it on our own.
5. The results process of our group project was tuff, but it really put everything together as a whole.
6. I think that your classes always challenge me to the extreme in this direction. In preparing for the proposal and presentation, reading many papers, trying to gain some sort of perspective on the field and where our research fit in. This was very challenging, but I feel like I have grown very much in this since the beginning of the semester.
7. –
8. –
9. Yes, for the same reason I put under "analysis."
10. –
11. Same as above, except in content of proposal project and poster.
12. Yes – every experience is helpful.
13. Research proposal and presentation really helped with this.
14. I think the presentation helped me the most because I have to really understand it to be able to talk about it.
15. Yes, the proposals helped develop this skill.

as they could have and may comment on unequal sharing of the workload. However, the group often feels that they made an effort to rectify these issues and as a result, learned from the experience and have an idea of how they would handle things differently in the future. These reflection pieces accomplish what they were intended to do, to make the students more mindful of the collaborative process that occurs in the sciences and how it requires effort on the part of all the participants for it to be a positive and productive experience.

GUIDED VERSUS OPEN-ENDED RESEARCH

Finally, I think it is important to address why I decided to use a research platform of my choosing rather than leaving it wide-open to the students to determine for themselves what they will investigate as their group projects. I think this choice was made in part from a sense of practicality because the reality of overseeing ten separate projects is logistically a nightmare and can detract from the research experience itself. But I think it is really more an issue of the undergraduates' developmental abilities and the lack of having had experienced previously a process-oriented approach to learning science. Not being familiar with a learn-by-doing approach can be intimidating when you do not have a frame of reference. That is not to say that the students do not have the intellectual ability to succeed with an open-ended format, but under the physical, financial, personnel, and time constraints that are operating in most undergraduate institutional contexts, a more guided experience offers the most beneficial outcome with the least amount of negativity and frustration on the part of the student and the faculty.

An additional layer to the choice of a research focus for the course is also institutionally contextualized but more from the lens of the faculty versus that of the student. My choosing to use my research program as the context for the students' projects is a little selfish on my part but from the broader perspective, directly speaks to my offering the students the best possible experience and an opportunity to be co-authors on a paper or conference presentation.

From a selfish perspective, I can advance my research agenda by having more time to spend on it since it is also a part of my teaching load. I have the flexibility to try out different research paths that might be more risky but have a greater transformational potential. Because these creative and less tried and true approaches are being worked on under the scope of a course, nothing is lost in the end. What may be gained is preliminary data that the students contributed that may redirect the research program's focus and/or lead to potential external funding.

The monetary issue is an important one considering financial constraints that exist on all campuses. The benefit of using a department's operational budget to fund a course that is part of the curricular program, and offers a means of helping to finance faculty scholarship (which is a requirement for tenure and promotion) can only be seen as a win-win scenario. Being able to use facilities, equipment, and expendables for two objectives simultaneously is not only prudent but a fiscally responsible means of making funds go further.

From a curricular focus, you are now providing a research experience to more of your students and not just to the high achievers or those who have already shown an interest in, and aptitude for, attending graduate school. If this is looked at from a pedagogical perspective, the value of doing science to learn about science should be made available to as many students as possible, science and non-science majors alike. Bringing authentic research experiences in to the classroom regardless of the disciplinary focus benefits the institution campus-wide. This parallels what the Council on Undergraduate Research (CUR) has been advocating for over 30 years. In a 2012 CUR publication, *Characteristics of Excellence in Undergraduate Research (COEUR)*, a summary of best practices includes reference to the integration of teaching and research, "... in courses teaching laboratory techniques in the sciences, it is often no more difficult to incorporate real research problems ..." (Characteristics of Excellence in Undergraduate Research [COEUR], 2012, p. 14). Supporting faculty as they collaborate with students on research, scholarly and creative activities benefits all parties and advances the mission of colleges and universities as well as higher education in general.

CONCLUSION

During the 23 years that I have taught undergraduates, I have seen an ongoing, and recently accelerating, movement from a teacher-centered classroom to a student-centered one. This pedagogical shift has resulted in beneficial student learning outcomes. All the classes that I teach, including the Genetics Lab described in this chapter, are structured on the student being actively engaged and participating in their learning. Science is by its very nature an active discipline that problem solves through observation and experimentation. Because it is evidence-based, it requires a hands-on means of obtaining reliable data. To not provide authentic research experiences for our students throughout the curriculum is a lost opportunity to help them identify with and become able practitioners of the scientific process. Too much is at stake.

I would go even further to say that exposing students to an authentic, inquiry and hypothesis-driven research experience is no longer just about the sciences. It is integral to every discipline that a student declares as a major and should be incorporated into all college curricula. All undergraduates should be knowledgeable, not only about what scholarship in their area of specialty looks like, but how to engage in the process. They should be prepared for their future career so that by the time they graduate they not only function as a practitioner, but as a competent scholar who can advance their discipline. They should see themselves as capable of contributing to future discoveries, innovations, and technical no how that can come to bear on finding solutions to the problems and challenges of the 21st century.

REFERENCES

Characteristics of Excellence in Undergraduate Research. (2012). Retrieved from http://www.cur.org/publications/publication_listings/COEUR/

Council on Undergraduate Research. (2011). *Fact sheet*. Retrieved from http://www.cur.org/about_cur/fact_sheet/

Hmelo-Silver, C. E. (2004). Problem-based learning: What and how do students learn? *Educational Psychology Review*, *16*(3), 235.

King, A. (1993). From sage on the stage to guide on the side. *College Teaching*, *41*(1), 30–35.

Kuh, G. D. (2008). *High impact educational practices: What they are, who has access to them and why they matter*. Washington, DC: Association of American Colleges and Universities.

Laursen, S. L., Hunter, E. S., Thiry, H., & Melton, G. (2010). *Undergraduate research in the sciences: Engaging students in real science*. New York, NY: Jossey-Bass of John Wiley & Sons, Inc.

Lopatto, D. (2003). The essential features of undergraduate research. *Council on Undergraduate Research Quarterly*, *23*(1), 139–142.

Lopatto, D. (2004). Survey of undergraduate research experiences (SURE): First Findings. *Cell Biology Education*, *3*(4), 270–277.

National Science Foundation. (1994). The Creation of an Undergraduate Molecular Genetics Course Based on Cooperative Learning. Grant No. DUE-9452376.

National Science Foundation. (1998). The Creation of an Integrated and Investigative Molecular and Cellular Biology Emphasis. Grant No. DUE-9851477.

Renkl, A., Atkinson, R. K., Maier, U. H., & Staley, R. (2002). From example study to problem solving: Smooth transitions help learning. *Journal of Experimental Education*, *70*(4), 293–315.

Vision and Change in Undergraduate Biology Education: A Call to Action. (2011). Retrieved from http://visionandchange.org/finalreport/

IBL AND MATHEMATICS: THE NECESSITY OF *FLEXIBLE* KNOWLEDGE

Vanessa R. Pitts Bannister and Gina J. Mariano

ABSTRACT

In this chapter, we discuss cases involving pre-service mathematics teachers. These cases highlight teachers' understanding of perspectives as they solve and analyze student work involving tasks that call for translations between algebraic and graphical representations. Findings suggest that flexible or inflexible knowledge structures influence participant's attempt to analyze or address student thinking.

Findings from this study highlight a major concern for teacher education and IBL efforts. If teachers do not have a flexible knowledge base to draw from, teachers may fail to capture important aspects of students' thinking and help students develop appropriate understandings. Such failure may result in missed opportunities to effectively assist students to explore, create and communicate, ideas that are core to IBL.

Inquiry-Based Learning for Science, Technology, Engineering, and Math (STEM) Programs: A Conceptual and Practical Resource for Educators
Innovations in Higher Education Teaching and Learning, Volume 4, 135–149

ISSN: 2055-3641/doi:10.1108/S2055-364120150000004009

Tell me and I forget, teach me and I may remember, involve me and I learn.

This often-used quote encompasses inquiry-based learning in that students are perceived as active participants in their own learning. According to The Academy of Inquiry Based Learning (2014), an instructor of Inquiry Based Learning (IBL) does not merely show a solution to a problem, but instead guides students in sense-making activities and tasks, which allow them to explore and create knowledge. Unlike IBL and in many mathematics classrooms, learning mathematics means following the rules and remembering and applying these rules (Ernst, 2013). Inquiry-based learning challenges students to think like mathematicians and acquire their own knowledge by creating/discovering mathematics (Ernst, 2013). Most important, IBL aligns with reform-oriented efforts, such as Common State Standards for Mathematics (CSSM), in that it promotes "knowledge of mathematics content and reasoning skills that enable students to apply mathematics effectively in a myriad of careers and in everyday life" (NCTM, 2013).

Teachers play an important role in learning, particularly, teachers must create opportunities for students to rediscover and reinvent concepts and methods of solution (Lappan, 1998). To be effective in this role, teachers should engage "students in meaningful learning through individual and collaborative experiences that promote their ability to make sense of mathematical ideas and reason mathematically" and consistently provide "students, individually and collectively, with opportunities and supports to engage in productive struggle as they grapple with mathematical ideas and relationships" (NCTM, 2014, p. 10). Moreover, teachers must exhibit high expectations (NCTM, 2014). High expectations can be exhibited in various ways. For instance, Jamar and Pitts (2004, p. 46, 2005, p. 130) provide the following guidelines for practice:

a) Using students' prior knowledge as building blocks to new knowledge;
b) Expecting students to be active participants in their own learning; and
c) Providing opportunities for students to *understand* concepts prior to learning rules.

In order to accentuate the above pedagogical elements of IBL, teachers engage in a continuous cycle of active listening to mathematical conceptions of students, showing support of mathematical conceptions, expressing concerns regarding conceptions, tracking ideas, and thought patterns in order to provide appropriate guidance.

As teachers undertake these facets (Fig. 1), they must demonstrate sound knowledge that highlights an interdependent relationship between

Fig. 1. Facets of *Constructive Opportunities* (Pitts Bannister, Mariano, & Hall, 2010).

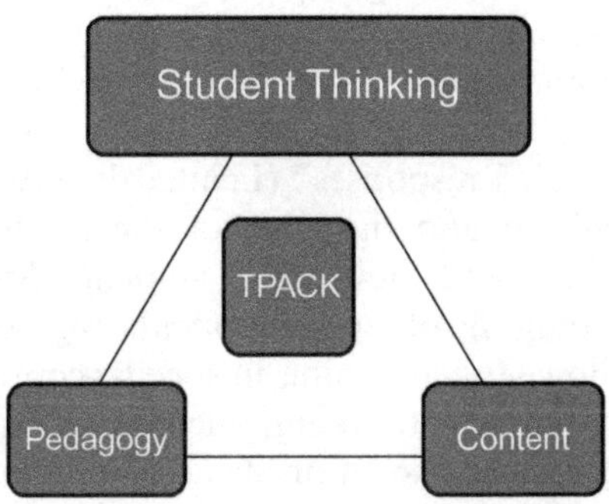

Fig. 2. Interdependence Model. *Source*: Pitts Bannister and Mariano (2013).

student thinking, pedagogy, and content knowledge. For instance, with respect to the development of Technology, Pedagogy and Content Knowledge (TPACK); (Mishra & Koehler, 2006; Niess et al., 2009) student thinking is paramount because it indicates the nature of knowledge related to content and pedagogy.

Most importantly, as illustrated in Fig. 2, TPACK arises within the context of the interplay between student thinking, pedagogy, and content. More specifically, the interactions between these key areas should inform pedagogical choices with respect to technology. Similarly, the interactions between student thinking, pedagogy, and content are paramount to IBL efforts because they should inform pedagogical choices with respect to inquiry and exploration.

In this chapter, we discuss cases involving pre-service mathematics teachers. These cases highlight the kinds of practices that may foster inquiry in the classroom. Particularly, this investigation emphasizes the role of teacher knowledge in effective implementation of IBL.

THE NECESSITY OF *FLEXIBLE* KNOWLEDGE

Teaching encompasses many facets of knowledge. Included among them is a cycle through activities of comprehension, transformation, instruction, evaluation, and reflection (Shulman, 1987). The core of these facets is comprehension for "to teach is first to understand" (Shulman, 1987, p. 14). In addition to this notion, "teaching requires the display of both static, organized knowledge bases and adaptive action systems of productions. Both static and dynamic understanding will be discussed throughout this paper. Further, in teaching, knowledge is used under conditions that change swiftly and demand varied responses" (Leinhardt, 1993, p. 48). Along with considering the cognition and thoughts of the teachers, Peterson (1988) emphasizes the cognition and thoughts of students. Specifically, Peterson's framework includes (among other components regarding teachers' cognition and thoughts): how students think in specific content areas and how to facilitate growth in students' thinking, such as increasing flexible understanding about a particular concept or idea.

The amalgamation of the above perspectives is effectively captured in the new *Principles to Actions: Ensuring Mathematical Success for All* document of the National Council of Teachers of Mathematics (NCTM), which focuses on preservice teacher education. Specifically, it identifies the following eight guidelines for effective teaching:

- Establishes clear goals for the mathematics that students are learning, situates goals within learning progressions, and uses the goals to guide instructional decisions.
- Engages students in solving and discussing tasks that promote mathematical reasoning and problem solving and that allow multiple entry points and varied solution strategies.
- Engages students in making connections among mathematical representations to deepen understanding of mathematics concepts and procedures and as tools for problem solving.
- Facilitates discourse among students to build shared understanding of mathematical ideas by analyzing and comparing student approaches and arguments.

- Uses purposeful questions to assess and advance students' reasoning and sense making about important mathematical ideas and relationships.
- Builds fluency with procedures on a foundation of conceptual understanding so that students, over time, become skillful in using procedures flexibly as they solve contextual and mathematical problems.
- Provides students, individually and collectively, with opportunities and supports to engage in productive struggle as they grapple with mathematical ideas and relationships.
- Uses evidence of student thinking to assess progress toward mathematical understanding and to adjust instruction continually in ways that support and extend learning. (NCTM, 2014, p. 10)

As briefly mentioned above, the cognition and thoughts of both teachers and students lay the groundwork for teaching. Accordingly, in order to support IBL efforts or guide students in sense-making activities, teachers must possess a sound knowledge base with a pedagogical basis (Darling-Hammond, 1998). Teachers must have a deep understanding of the material they are teaching in order to promote inquiry in the classroom.

Knowledge Transfer and Translations

Researchers in the area of learning have studied and supported the concept of knowledge transfer and its importance in academic settings for decades (Mariano, 2014). Knowledge transfer can be defined as the ability to apply knowledge or use knowledge from one problem, situation or context to another (Anderson, 2005). For instance, near transfer, or specific transfer, refers to the transfer that occurs between two situations or tasks that are similar in both their superficial and underlying characteristics and principles (Mariano, 2014). Zero transfer occurs when learning in one situation has no effect on learning in another situation, meaning there is no flexible movement between multiple perspectives (Mariano, 2014). Concepts of knowledge transfer can be associated with notions of translations between algebraic and graphical representations. For instance, the extent to which individuals move flexibly between these representations relies on their ability to understand the correspondence between the representations or flexibly operate in process or object perspectives. Specifically in mathematics, translations involve the ability to perform at least one of the following:

- Identify a specific transformation of a function in one representation and its corresponding transformation in another representation (Leinhardt, Zaslavsky, & Stein, 1990);

- Recognize the same function in different forms of representations (Leinhardt et al., 1990);
- Generate one representation of a function given another one (Leinhardt et al., 1990); or
- Illustrate an understanding that a point is on a graph of a line if and only if its coordinates satisfy the equation of the line (Moschkovich, Schoenfeld, & Arcavi, 1993).

To highlight such proficiencies let's consider a problem (Fig. 3) administered to 12th grade students during the Sixth Mathematics Assessment of the National Assessment of Educational Progress (NAEP).

Blume and Heckman (1992) suggest that only 20% of the 12th grade students could effectively complete the above problem. As one attempts this problem, some questions to consider include: how do the output values (or y-values) change as the function $y = f(x)$ is changed to $y = |f(x)|$, or how does the change of $y = f(x)$ to $y = |f(x)|$ affect the graphical representation of $y = f(x)$? As such questions are addressed, one uses information within the algebraic representation to determine its effect with respect to the graphical representation and vice versa. In order to perform these tasks, one must flexibly operate within two perspectives: process and object (Moschkovich et al., 1993). The process perspective is to highlight the relationship between the x and y values of a functional relation, while the

A NAEP Exercise

The following graph represents y = f(x).

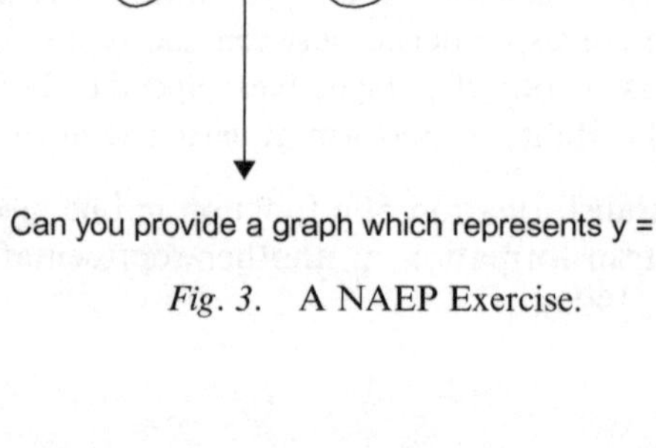

Can you provide a graph which represents y = |f(x)|?

Fig. 3. A NAEP Exercise.

object perspective is to associate values within a functional relation to movements or positioning of a graph.

METHODS

Participants

The data to be discussed were part of a larger study (see Pitts Bannister, 2014) involving 59 prospective mathematics teachers from four northeastern universities in the United States. Participants of the larger study were seniors in a teacher education program or students in a professional certification program. The students were solicited from a required mathematics methods course. In addition, these students were going to be student-teachers during the upcoming fall or spring semester. This group of participants was preferred so that the description of their knowledge would reflect knowledge teachers have gained during their college education, but before they start teaching.

Data Collection and Sources

Data for this study were collected in two phases. In the first phase, 59 prospective mathematics teachers completed a two-part questionnaire (parts A and B). Questionnaire A included six open-ended problems, which were designed to determine the manner in which participants perform problems that call for translations between algebraic and graphical representations of functions. The problems were adopted from several sources (Blume & Heckman, 1992; Knuth, 2000; Moschkovich et al., 1993). In order to solve the problems, participants had to effectively perform translations between algebraic and graphical representations of functions and operate within both process and object perspectives. For each problem, the participants were asked to solve a problem and explain their solution methods. In the second phase, 10 participants agreed to be interviewed to further describe their solution methods presented during phase 1 and address additional appropriate questions.

For the purpose of this study, we will focus on the following four problems (Fig. 4), which were adopted from Moschkovich et al. (1993).

Translation Tasks

Using the following graph, solve problems 1–4. Please show all work.

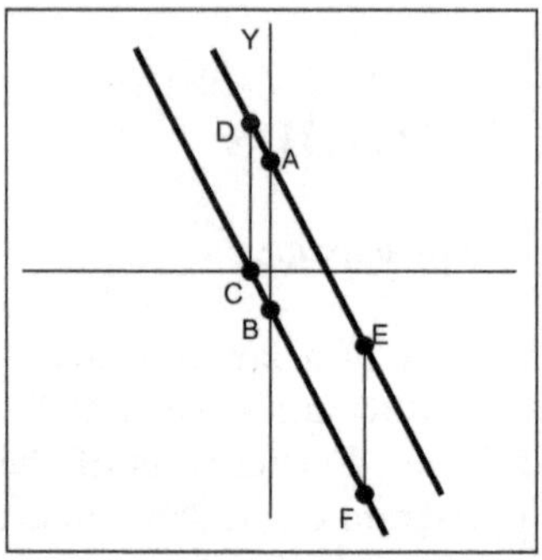

1) The following list includes the equations of the two lines. Match each line with its equation.
$y = 2x + 6$ $y = 2x - 2$ $y = -2x - 2$ $y = -2x + 6$

2) Find the coordinates of points A, B, C, and D, knowing that the line segments CD and EF are parallel to the y-axis.
3) If the *x*-coordinate of point E is 5, find its *y*-coordinate and the coordinates of point F.
4) Find the lengths of segments EF, CD, and AB.

Fig. 4. Translation Tasks. *Source*: Adapted from Moschkovich et al. (1993).

Dynamic and Static Understandings

Two participants were chosen to highlight some of the differences in flexibility among prospective teachers' understandings. One of the participants exhibited a static understanding, whereas the other participant exhibited a dynamic understanding. Static understanding is defined as an individual's failure to move between multiple perspectives while solving problems that call for translations. Whereas, by contrast, dynamic understanding is defined as an individual's ability move flexibly between multiple perspectives while solving problems that call for translations. This is when knowledge transfer is occurring.

Asia's Approach

Below is Asia's response to problem 1 (Fig. 5).

Asia provided the following explanation of her solution method:

Asia's Work

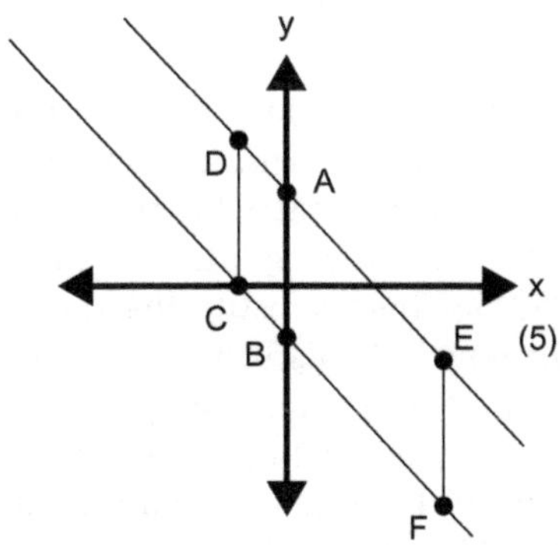

(1) The following list includes the equations of the two lines. Match each line with its equation

$y = 2x + 6$ $y = 2x - 2$ $y = -2x - 2$ $y = -2x + 6$

Please explain how you obtained your answers

BF$\Rightarrow y = -2x - 2$

Subtituting values for *x*, you get the values for *y*.

eg.

y	*x*
−2	0
−4	1
−6	2
−8	3

y	*x*
0	−1
−12	5

Same is the case for AE$\Rightarrow y = -2x + 6$

eg.

y	*x*
6	0
4	1
2	2
0	3

y	*x*
−2	4
−4	5
8	−1

Fig. 5. Asia's Response to Problem 1.

> I did a table of values. From the table of values I found out from the various equations that's given on top. Substituted it and found the one that would match the line with the equation.

Asia used the above table of values to determine the coordinates for points A, B, C and D for problem 2. Specifically, she stated, "the table helped me look for it [coordinates for points A, B, C and D]". When asked if she could provide another solution method for problem 2, Asia stated the following: "Other than substituting the values for 'x' and 'y'? I can't think of one right now." Asia's solution methods for problems 1 and 2 heavily relied on constructs of process perspective. In particular, her solution methods emphasized an understanding of the *Cartesian Connection* – "[a] point is on the graph of the line L if and only if its coordinates satisfy the equation of L" (Moschkovich et al., 1993, p. 73). Although Asia's solution methods embodied this level of understanding, it is limited to process perspective. In light of this observation, Asia demonstrated a static understanding. Most important, Asia's static understanding seemed to hinder her attempt to identify constructs of object perspective as she analyzed the following student response, which was based on problem 2.

> Student Response: "A equals (0, A) and B equals (0, –B)"

Asia presented the following analysis of the above response:

> The student is thinking in terms of the alphabet and the coordinate point. The student is looking at the graph and just writing down the point and the 'x' or 'y' point with it.

During the interview, Asia further explained this analysis by stating the following:

> Probably the student is just looking at A equals y and –B equals y. I think the student is just looking at the graph and has jotted down whatever she sees.

Based on Asia's analysis of the student response, she viewed the student's misconception as a mere mislabeling of points. But she may have missed important information about the student's thoughts. For instance, it was clear that the student recognized points A and B as y-intercepts by noting the x-coordinates of each point as 0. Moreover, the student acknowledged the locations of the points, with respect to the y-axis, by noting their y-coordinates as –A and B. Asia did not acknowledge these aspects, which embodied constructs of object perspective. More importantly, Asia did not move beyond constructs of process perspective to analyze the student's response. She continued to solely rely on constructs of process perspective as she attempted to address the student's misconception. Asia presented the following:

> Explain to the student that a point has two coordinates' values- one for 'x' and one for 'y'. These two values should satisfy the equation. (Help with plug and chug).

She further explained her pedagogical approach by stating the following:

> If the line has an equation that goes with it, help the student determine like maybe two positives [x- and y-] values, two negative [x- and y-] values and (0,0)… and see if that satisfies the equation.

Kyanne's Approach

After successfully completing the problems, Kyanne stated the following regarding her solution method for problem 1:

> I guess you could say it was more or less eliminating the ones that could not be correct and then identifying … then what that [the equations] must be from that. So, I looked at the two lines and saw they had a negative slope. Knowing the $y = mx + b$ form, that eliminated two [equations] that had a positive slope. And then for line ED I said well the y-intercept is above the x-axis, which is a positive y-value and that gave me $y = -2x + 6$. Since there was only one [equation] left that had a negative slope that one would have to be AC.

From the above information and using the components of the relation for a straight line Kyanne determined the coordinates of points A and B. In order to determine point C, Kyanne chose substitution methods. For instance, Kyanne stated the following:

> If we assume that CD is vertical then we can find the value of C using the equation for the line and knowing that at the x-intercept the value of y is zero. So plugging in y equals zero we can solve for x … x is negative 1. So, C coordinates are (−1, 0).

Kyanne's solution methods involved constructs of process and object perspectives. In particular, when determining the coordinates of points A and B, Kyanne demonstrated an understanding of the components of the functional relation, $y = mx + b$ (where m is the slope, and b is the y-intercept). This approach highlighted constructs of object perspective. Kyanne introduced constructs of process perspective as she determined the coordinates of point C by substituting 0 into the appropriate equation. Kyanne's solution methods for problems 1 and 2 demonstrated a dynamic understanding, which was characterized by her demonstration of constructs of process and object perspectives. Kyanne's understanding of the problem seemed to play a major role in her attempt to analyze student response 1. For instance, Kyanne presented the following analysis:

> The student is not considering the equations of the lines. They looked at the point A and the point B and interpreted A, B as vertical distances from the origin. This is probably the result of the student seeing units along the axes and confusing units with points. My initial response is that is not correct. However, it could be considered correct if the student indicated that the height of A above the origin is A, and the distance below the origin of B is B.

Kyanne further explained this analysis by stating the following:

> So I think what ends up happening is that a student looks at a point, looks at the point A and says: well it's 'a' units above the x-axis and a y-intercept, so (0, A). But then the student says well … this one is 'b' units below [the x-axis] and really should say negative (0, −B).

Another important aspect of Kyanne's dynamic understanding was her ability to analyze the student response and attempt to assist the student via constructs of object perspective. Kyanne presented the following explanation as to how she would address the student's misconception(s) concerning problem 2:

> Again, it may not be that the student misunderstands the graph as much as how units and points are used on the graph. I would be more concerned that the student did not realize that by plugging $x = 0$ into the equations, they could find numerical answers for the y-values. That is where I would focus my assistance.

Kyanne's ability to move back and forth between perspectives while analyzing and attempting to address one student's misconception could be attributed to her dynamic understanding demonstrated within her solution methods for problems 1 and 2.

DISCUSSION

Teachers play an important role in learning as they engage in a continuous cycle of active listening to mathematical conceptions of students, showing support of mathematical conceptions, expressing concerns regarding conceptions, tracking ideas, and thought patterns in order to provide appropriate guidance (Pitts Bannister et al., 2010). Asia's and Kyanne's responses highlight a major concern for teacher education programs, particularly how they relate to effective implementation of IBL. If teachers do not have a flexible knowledge base to draw from, teachers may fail to capture important aspects of students' thinking and help students develop appropriate understandings. This may limit student opportunities for inquiry. Hence, limited content knowledge can be a hindrance. For instance, the fact that Asia did not respond to all aspects of the student response, which embodied constructs that differed from her knowledge base, could be attributed to her static understanding. In other words, this finding highlights the possible limitations of static understanding in that one may not be positively positioned to support IBL efforts or guide students in sense-making activities. By contrast, Kyanne, who exhibited a dynamic understanding, adequately identified and utilized constructs of object perspective to analyze and address the student's misconception. The eight NCTM guidelines listed at the beginning of this chapter can be used to encourage IBL in the mathematics classroom. Through the use of purposeful questions to assess and advance students' reasoning and sense making about important mathematical ideas and relationships, student can begin to build flexible understanding. Teachers education programs can design environments that continually support and extend learning through the use of inquiry.

To conclude this paper, it is worth repeating the opening quote of this paper: *Tell me and I forget, teach me and I may remember, involve me and I learn*. This quote captures the essence of this paper in that the cases provide examples of each idea within the quote. For instance, Asia's approach only emphasized the first two ideas of the quote in that it did not incorporate (or involve) the constructs of the student's conception. Kyanne's approach was different in that it incorporated (or involved) the student's conceptions. The variance in approaches may result in missed opportunities to effectively assist students to explore, create, and communicate, ideas that are core to IBL.

REFERENCES

Anderson, J. R. (2005). *Cognitive psychology and its implications*. New York, NY: Worth.

Bannister, V. R. P. (2014). Flexible conceptions of perspectives and representations: An examination of pre-service mathematics teachers' knowledge. *International Journal of Education in Mathematics, Science and Technology*, *2*(3), 223–233.

Blume, G. W., & Heckman, D. S. (1992). What do students know about algebra and functions? In P. A. Kenney & E. A. Silver (Eds.), *Results from the sixth mathematics assessment* (pp. 225–277). Reston, VA: National Council of Teachers of Mathematics.

Darling-Hammond, L. (1998). Teachers and teaching: Testing policy hypotheses from a national commission report. *Educational Researcher*, 27, 5–15.

Ernst, D. (2013, May 29). *What the heck is IBL?* Retrieved from http://maamathedmatters.blogspot.com/2013/05/what-heck-is-ibl.html

Jamar, I., & Pitts, V. R. (2004). Bringing high expectations to life in an urban classroom. In M. Chappelle & T. Pateracki (Eds.), *Empowering the beginning teachers of mathematics in middle school.* Reston, VA: NCTM.

Jamar, I., & Pitts, V. R. (2005). High expectations: A "how" of achieving equitable mathematics classrooms. *Negro Educational Review*, *56*(2–3), 127.

Knuth, E. (2000). Student understanding of the Cartesian connection: An exploratory study. *Journal of Research in Mathematics Education, 31*, 500–508.

Lappan, G. (1998). Pedagogical Implication for Problem-Centered Teaching. In *Mathematics sciences education board, high school mathematics at work: Essays and examples for the education of all students* (pp. 132–140). Washington, DC: National Academy Press.

Leinhardt, G. (1993). 'On Teaching'. In R. Glaser (Ed.), *Advances in instructional psychology* (Vol. 4, pp. 1–54). Hillsdale, NJ: Erlbaum.

Leinhardt, G., Zaslavsky, O., & Stein, M. (1990). Functions, graphs, and graphing: Tasks, learning, and teaching. *Review of Educational Research*, *60*, 1–64.

Mariano, G. J. (2014). Breaking it down: Knowledge transfer in a multimedia learning environment. *International Journal of Teaching and Learning in Higher Education*, *26*(1), 1–11.

Mishra, P., & Koehler, M. J. (2006). Technological pedagogical content knowledge: A framework for teacher knowledge. *Teachers College Record*, *108*(6), 1017–1054.

Moschkovich, J., Schoenfeld, A. H., & Arcavi, A. (1993). Aspects of understanding: On multiple perspectives and representations of linear relations and connections among them. In *Integrating research on the graphical representation of functions* (pp. 69–100). Hillsdale, NJ: Lawrence Erlbaum Associates.

National Council of Teachers of Math-ematics (NCTM). (2014). *Principles to actions: Ensuring mathematical success for all.* Reston, VA: NCTM.

National Council of Teachers of Mathematics (NCTM). (2013). *Supporting the common core State standards for mathematics.* Reston, VA: NCTM.

Niess, M. L., Ronau, R. N., Shafer, K. G., Driskell, S. O., Harper, S. R., Johnston, C., … Kersaint, G. (2009). Mathematics teacher TPACK standards and development model. *Contemporary Issues in Technology and Teacher Education*, *9*(1), 4–24.

Peterson, P. (1988). Teachers' and students' cognitional knowledge for classroom teaching and learning. *Educational Researcher*, *17*, 5–14.

Pitts Bannister, V. R., & Mariano, G. J. (2013). Snapshots of student thinking: An exploration of video cases for extending prospective teachers' knowledge. In C. Craig & N. Gallavan (Eds.), *Action in teacher education yearbook XXII* (Vol. 35, pp. 430–444). New York, NY: Routledge.

Pitts Bannister, V. R., Mariano, G. J., & Hall, C. D. (2010). Creating constructive opportunities: A "how" to embracing students' mathematical conceptions. In M. Q. Foote (Ed.), *Mathematics teaching and learning in K-12: Equity and professional development* (pp. 167–180). New York, NY: Palgrave.

Shulman, L. S. (1987). Knowledge and teaching: Foundations of the new reform. *Harvard Educational Review*, *57*(1), 1–23.

The Academy of Inquiry-Based Learning. (2014). *What is IBL?* Retrieved from http://www.inquirybasedlearning.org/?page=What_is_IBL

ADVANTAGES OF SCIENCE EDUCATION OUTDOORS THROUGH IBSE METHODOLOGY

Ana Cristina Tavares, Susana Silva and Teresa Bettencourt

ABSTRACT

Science Education Outdoors provides students with direct contact with natural phenomena and enables active learning, a key factor in Inquiry Based Science Education (IBSE), a student-centred methodology for the acquisition, construction and understanding of knowledge.

This chapter will describe three case studies which used the IBSE methodology as both a teaching and learning methodology, promoting a deeper understanding of how IBSE can contribute to the success of learning and teaching in outdoor settings.

The three case studies were based on three training courses conducted at the Botanical Garden of the University of Coimbra (BGUC, Portugal). The first case study was the annual regular course for garden educators, and the other two were the two editions of the COInquire professional and accreditation training course for teachers and educators. Involving a

Inquiry-Based Learning for Science, Technology, Engineering, and Math (STEM) Programs: A Conceptual and Practical Resource for Educators
Innovations in Higher Education Teaching and Learning, Volume 4, 151–169

ISSN: 2055-3641/doi:10.1108/S2055-364120150000004010

total of 70 participants, data was collected through the application of questionnaires.

The study revealed that all participants considered IBSE a successful teaching–learning process and they remarked the opportunities created for the active construction of new knowledge. Strengthened by numerous live educational resources, the use of IBSE in the garden facilitated the questioning and interpretation of nature, supporting the open-minded and well-founded training of teachers, educators and students.

Additionally, the participants considered IBSE to be an effective methodology to boost their professional improvement, contributing to the development of innovative approaches to the curricular programmes on biodiversity and sustainability.

INTRODUCTION

Science Education Outside the Classroom – A Brief Framework

Every life experience has a formative effect. One of the most important contributions resides in what and how we learn through formal, non-formal or informal scenarios, being that the adults and children's specific interest in science is encouraged by learning environments outside the classroom, namely botanical gardens, museums or simply the school garden, which are very rich in stimuli and educational resources (ASE OSWG, 2011; Bora et al., 2010; Gatt, Tunnicliffe, Borg, & Lautier, 2007; Kapelari, Vergou, Willison, Regan, & Dillon, 2014; National Research Council, 2009; Rudd, 2008; Tavares, 2015a).

> Children who have the freedom to explore wastelands and woodlands … will all be 'discoverers' and too little of this is possible in formal schooling. The over-sheltered child must often be the poorer in their learning. (Tomkins & Tunnicliffe, 2007, p. 155)

Science and Nature activities performed outside the classroom, as is the case of educational programmes developed in Botanical Gardens, are an excellent instructive practice, facilitating the knowledge acquisition process. As they have a living space as study environment, educational activities in the Botanical Garden are inevitably dynamic and very enriching, in addition to offering an immense availability of educational resources and direct contact with real models, authentic examples of natural phenomena, which are engaging and often surprising (Tavares, 2015a). 'Therefore, real life, out-of-school experiences are paramount to learn about organisms' (Patrick & Tunnicliffe, 2011, p. 631).

There is a motivational value in the observation of natural objects and their positive effect on attitudes, as well as their usefulness in developing children's knowledge of natural phenomena (Patrick & Tunnicliffe, 2011; Tomkins & Tunnicliffe, 2007). Tunnicliffe and Ueckert (2007, p. 52) argue that biologists need to reconnect students with the natural world *first hand,* so as to give them the opportunity to identify natural history – the study of organisms in their natural environment – as the soul of Biology (Page, 2005, pp. 361–362), taking into account that contextual and cultural examples are pertinent in biology teaching, because habitats and species vary greatly, even within the same country (Patrick et al., 2013; Patrick & Tunnicliffe, 2011).

For the interpretation of non-formal educational settings, namely natural history and science museums and botanical gardens, the explainers – general designation for the monitors (guides, teachers or educators) of these spaces – are key communicators and promoters of knowledge relevant to achieve an effective teaching–learning process.

Explainers are 'issue specialists' and the first ones meeting visitors at these science institutions. They must be well prepared, welcoming and helpful, in addition to having the needed information, and being able to inspire and empower visitors, identifying and answering their questions, as well as sharing their fascination for knowledge (Exploratorium, 2014; Science Museum, 2014; Scientix European Plataform, 2014; Tavares & Silva, 2014).

These institutions outside the classroom, such as gardens, are rich, unique and extremely inspiring. Literally speaking: people get more easily inspired outdoors – there is more oxygen – in stimulating environments which are tailored to focus on each visitor's specific interests and, in turn, will also enhance the group's interests, promoting knowledge on material heritage by stimulating and empowering each person's immaterial identities as well as that of the entire group (ASE OSWG, 2011; Bora et al., 2010; Bromely et al., 2013; National Research Council, 2009; Rudd, 2008).

When visiting spaces outdoors, experiences arise which have a special meaning and effect on very young visitors, as they are apprentices, beginners. This is the perfect time to combine children's first thoughts and reflections about nature with the importance of plants for the sustainability of life (Bora et al., 2010; Gatt et al., 2007; Rudd, 2008; Tavares et al., 2014b).

Young children are spontaneously active and inquisitive 'by nature'. They are innately interested in all living things, and have a passion for understanding and exploring their immediate surroundings, both inside and outside their home (Gatt et al., 2007).

Parents have been identified as the children's main source of knowledge at home in the early years, so school educators and teachers need to focus on the development of ways for children to observe and look for meaning in nature, particularly plants which are often within reach (Gatt et al., 2007; Patrick & Tunnicliffe, 2011; Tavares, 2014).

In addition to the assortment of plant collections, which are mostly unique, Botanical Gardens hold other living, historic and architectonic heritage which is available and can help the public engage with natural science and other kinds of knowledge, enhancing general literacy. Functioning as a theatre, laboratory and house of life, Botanical Gardens are particularly empowering settings, and can inspire new artistic, cultural and educational research or merely leisure undertakings. It is a waste not to enjoy them (Tavares, 2015b).

Yet, during a visit to a site outdoors, such as botanical gardens, there is a large amount of general or specific teaching–learning resources and tools at hand, and further educational questions, innovations and challenges are expected for teachers, educators, and learners.

What Is IBSE?

Currently, there seems to be some consensus among the scientific community dedicated to educational research, regarding the fact that inquiry is the essence of science education, and that student-directed inquiry, in educational programmes, aids the development of a more competent professional (Bora et al., 2010; Cachapuz, Praia, & Jorge, 2002; Dillon, Elster, & Kapelari, 2011; Kapelari et al., 2012, 2014; Lopes & Bettencourt, 2012; Patrick et al., 2013). Contrary to teachers merely introducing scientific content, the IBSE methodology focuses on student inquiry, perceived as the driving force for science learning, being that teaching is organised around questions and problems within a student-centred inquiry process (Pathway Educative European Project, 2011).

The inquiry learning environment promotes an active, student-centred and student-led learning approach, based on constructivist learning theories, which incorporates the students' existing preconceptions and tries to support conceptual reconstruction processes (Duit & Treagust, 2003).

There are several educational definitions of inquiry. Following, we present some to be taken into consideration:

- 'Scientific inquiry requires identification of assumptions, use of critical and logical thinking, and consideration of different explanations by finding answers to questions' (National Research Council, 1995, cit. in Bromley et al., 2013, p. 9).
- Inquiry can be defined as 'The intentional process of diagnosing problems, critiquing experiments and distinguishing alternatives, planning research projects, researching conjectures, searching for information, constructing models, debating with peers, and developing coherent arguments' (Linn, Davis, & Bell, 2004).
- 'Inquiry-based science education means students progressively developing key scientific ideas through learning how to research and build their knowledge and understanding of the world around them. They use skills already used by scientists such as raising questions, collecting data, reasoning and reviewing evidence in light of what is already known, drawing conclusions and discussing results. Therefore, inquiry-based teaching leads students to build their understanding of fundamental scientific ideas through direct experience with materials, by consulting books, other resources, and experts, and through argument and debate among themselves. All this takes place under the leadership of the classroom teacher' (Wynne Harlen, 2011, cit. in Bromley et al., 2013, p. 10).

To implement an enquiry-based methodology there are several educational strategies, such as the one described in the European Pathway Project, which is structured around seven essential features: (1) Question – Students research scientifically-oriented questions; (2) Evidence – Students give priority to evidence; (3) Analysis – Students analyse evidence; (4) Explain – Students formulate an explanation based on evidence; (5) Connect – Students connect explanations to scientific knowledge; (6) Communicate – Students communicate with an audience and justify their explanation; and (7) Reflect – Students reflect on the inquiry process, reaction to their work, its value and impact, and their learning.

Another strategy, used in the two Coimbra Inquire (COInquire) training courses and described in this chapter, follows five steps: (1) Identification of problem area/question; (2) Collection and organisation of data; (3) Interpretation of data; (4) Action based on data; and (5) Reflection.

In general, in the IBSE methodology, three main steps are identified:

1. Focus on the question/problem at hand, which requires the mobilisation of previous knowledge and the construction of new knowledge, allowing for the resolution of the question/problem.

2. Design, including multidisciplinarity, to enable recognition of the biosphere, to understand the world in its entirety and complexity. A teacher or educator must summarise and prepare, in order to develop a communication plan with methodological and assessment strategies, not forgetting the approach to everyday-life situations with methodological pluralism; the trainer plays a key role as organiser and promoter of a working environment conducive to the building of students' knowledge, in addition to helping them develop critical thinking.
3. The educational evaluation of student learning outcomes should be regulated and guided, leading to reflection and assessment, irrespective of reaching an appropriate response to the question-problem or how the process was developed and knowledge was achieved (Bromley et al., 2013).

Inquiry-based science education arises from hands-on scientific experiments carried out in the classroom, although active learning entails much more than the mere manipulation of materials, as students must be encouraged to discuss their results, connect them to present and previous researches, having time to process the information and associate it with previous knowledge (Wenning, 2005).

For students to succeed in today's workforce they must know the natural world and much more. Expensive equipment, PowerPoint slides and worksheets are not essential in the pursuit of good biology. Outdoor settings provide a wide range of opportunities to deliver high quality IBSE case studies (Patrick et al., 2013; Tavares, 2013b; Tunnicliffe & Ueckert, 2007).

Teachers and educators must acknowledge their students' previous knowledge and misconceptions, and restructure their teaching accordingly, as they do not come to class without some sort of previous knowledge – they have and bring personal experiences, despite the fact that their preconceptions on nature are not always correct (Patrick & Tunnicliffe, 2011).

The European INQUIRE project (IEEP, 2013) promoted training courses using settings outside the classroom and the IBSE concept was based on four major principles guiding the learning process: 'Principle 1: Experience first what should be applied later; Principle 2: Collect evidence to improve your practice; Principle 3: Take your time learning; Principle 4: Communicate results' (Kapelari et al., 2014, p. 275).

IBSE activities can use extremely simple scientific language, as 'The algae that wanted to be a flower', on plant evolution which is primarily tailored to attract 3–5-year-old children so as to awake interest in plant existence. This lesson can be done and applied either in botanical gardens (non-formal and informal settings) or elsewhere (formal settings, such as the classroom).

This IBSE activity is meant to encourage children to construct elementary botany knowledge and to acknowledge that plants are living organisms which are important to ensure Life on Earth. In the case under study, the activity ought to be very simple, easy and pleasant, including all the key requirements to get young children's attention, access their world, knowledge and interests in addition to trying to behave and think like them.

New cycles of IBSE teaching–learning processes can be constructed and developed to include new tasks and multidisciplinary knowledge in addition to enhancing skills acquisition and development, encouraging students to build learning by promoting the reconstruction of personal concepts alongside with a change in attitudes and values (Tavares, 2013a, 2014).

Although learner-centred, the IBSE method requires trainers to be well prepared, confident, knowledgeable and sure about how and when to use educational resources, organising a proper student learning environment.

The IBSE activities are open to an autonomous orientation of the student, what increases the possibility of him to produce interpretations and outcomes unexpected for the teacher. Accordingly, it requires good preparation, and prompt, efficient teacher guidance and reaction expanding the teacher and student's universes, backgrounds and competences.

Knowledge is naturally constructed by a student, according to evidence (Tavares, 2015a; Tavares, Silva, & Bettencourt, 2015). This increases the student's underlying creativity and interpretations that should require the trainer's assistance and guidance whenever needed. IBSE values pre-existing knowledge and a new open and team/social perspective of teaching where both teacher and student's individualities are taken into account, respecting their uniqueness.

Considering 'the responsibility of being a teacher', Pessoa (2013, pp. 29–30) enhances teachers' professional as well as personal talents when she refers that 'in each teacher there is also a person, keeping in mind the importance of personal aptitudes in the teaching/learning practice', adding that 'training programmes should comprise a well grounded balance between the cognitive and emotional dimensions'. In order to better develop educational activities, it is important that both teacher and educator gather all their skills, memories and competences, linking and empowering the emotional and intellectual knowledge arising from personal experiences (Tavares, 2015a).

With the objective of fulfilling an institutional mission, the BGUC' Educative Service was launched in October 1997 and was implemented until November 2013, presenting an annual programme, mainly focused on Natural Sciences education, and especially devoted to the promotion of curricular programmes using the IBSE methodology, applied in formal and non-formal contexts for teachers, educators and students (Tavares, 2015a).

Offering an Educational Programme with 10 permanent and about 20 seasonal topics (Tavares, 2008, 2015a), the BGUC' Educative Service had to implement the training of garden educators. Participants of different high-level courses were offered a structured basic training course (20 hours of lessons in the garden + individual work + final evaluation) on the different topics included in the regular training course for garden educators (Tavares & Silva, 2014). This course is the first case study described in this chapter.

Bracing the IBSE methodology, through the Inquire European Educative Project (IEEP, 2013), IBSE training courses for teachers and educators were also taken for three years (2011–2013) at the BGUC (Silva, Tavares, & Bettencourt, 2013; Tavares & Bettencourt, 2013; Tavares, Silva, & Bettencourt, 2013; Tavares et al., 2014a, 2014b, 2015). In this chapter we describe the two training courses for teachers and educators as our second and third case studies.

Next, the structure of the three courses as well as the methodology and the results will be described in detail.

METHODOLOGICAL TOOLS

Regular Garden Educators' Training – 1st Case Study

Throughout the years, resorting to the IBSE methodology, the garden educators' training courses at the BGUC resulted in an increase in the number of courses and valences of the Educational Programme to meet the requests of other schools, attracting new audiences and delivering different topics. Manuals, pamphlets and paths, illustrated books and booklets, guides, different materials and education kits were produced for visitors, students, teachers, and educators (Bromley et al., 2013; Tavares, 2013a, 2015a, 2015b).

The course analysed in our case study was held in 2012, and included 28 participants, who answered the assessment questionnaire at the end of the course, in order to assess the effectiveness of the IBSE methodology and their degree of satisfaction.

COInquire Courses – Professional Training for Teacher and Educators – 2nd and 3rd case studies

The Coimbra Inquire (COInquire) training course (CIEEP, 2014) 'Project INQUIRE: sustainability and biodiversity training' for teachers and educators is accredited by the Portuguese Scientific National Committee, allowing geography, geology and biology teachers to progress in their career. From the 42 participants who attended the two editions of the course, the majority were teachers of the 3rd Cycle of Basic Education and Secondary Education teachers, mostly Biology, Geography and Geology teachers (Silva et al., 2013; Tavares & Bettencourt, 2013).

The two editions of the course (2012, 2013) were completed in 60 hours: 30 hours were divided in five classes held in the botanical gardens (5×6 h) and the remaining 30 hours were spent doing autonomous work. The participants had to develop individual projects inside and outside the classroom for their students, applying the IBSE methodology (Bromley et al., 2013).

The same questionnaire was used in the two COInquire course editions, namely before and after the course. The goal was to assess the effectiveness of the IBSE methodology and to understand the main changes arising in the participants' attitude towards IBSE and the use of botanical gardens as well as other spaces outside the classroom as educational resources to be used with their students.

RESULTS

Following, the evidence, results, outcomes and reflections on IBSE outdoor training courses will be analysed.

Regular Garden Educators' Training

The participants revealed that the main motivation for attending the course was 'Deepening knowledge' followed by 'Increasing general culture' and 'Knowledge of the Botanical Garden' heritage' (Table 1).

In general, all items were evaluated positively. Regarding the objectives of the training, emphasis was on 'Up-to-date content relevance', 'Adequacy of objectives to the content addressed' and 'Level of initial expectations', demonstrating how important and up-to-date this action is. Regarding

Table 1. Relevance of Learners' Motivation to Attend this Training.

Relevance of Learners' Motivation to Attend this Training										
Item	Mo	Mín.	Máx.	%1	%2	%3	%4	%5	%6	%7
Increased general culture	2	7	1	23.1	30.8	11.5	19.2	11.5	0.0	3.8
Deepening knowledge	1	6	1	26.9	19.2	38.5	3.8	7.7	3.8	0.0
Updating knowledge	4	7	1	7.7	15.4	15.4	23.1	15.4	11.5	11.5
Knowledge of the Botanical Garden' heritage	4	7	1	23.1	15.4	3.8	30.8	11.5	11.5	3.8
Improvement of practical communication	6/7	7	1	3.8	3.8	7.7	11.5	11.5	30.8	30.8
Exchange experiences with other colleagues	6	7	2	0.0	3.8	11.5	7.7	19.2	30.8	26.9
Curiosity about the topic	5	7	1	15.4	11.5	11.5	7.7	23.1	11.5	19.2

Source: Tavares and Silva (2014).

trainer performance, the trainees highlighted 'Scientific knowledge', 'Availability to explain doubts' and 'Clarity of language'. Interpersonal relations concerning 'Human Relationship between trainees and trainers' and among trainees themselves were also evaluated very positively.

Professional Training for Teachers and Educators – COInquire Courses

Out of the 42 attendees of the two COInquire course editions, only 35 participants completed the course, which corresponds to a completion rate of 83.3%. All of the participants achieved the main goals of the course through the development of innovative and reproducible outdoor IBSE projects with relevant and diverse curricular topics, including biodiversity, climatic changes, and sustainability (Silva et al., 2013; Tavares & Bettencourt, 2013). The Botanical Garden of Coimbra and Sustainability topics were highlighted by 75% and 68% of participants, in case study 1 and 2 respectively, as areas where they learned the most (Tavares & Silva, 2014).

The outcomes and new resources included in the participants' portfolios are available in the COInquire platforms, and were disseminated through a National Inquire Forum organised in two editions (2012 and 2013) (Tavares et al., 2013, 2014a). As seen from the projects developed, the participants' performance as teachers was successful and they used spaces outdoors, such as Botanical Gardens, for the development of the teaching and learning process in Science.

In fact, most of the participants, 40% in the first course and 31% in the second course, resorted to more than one outdoor space, highlighting not only the use but also the type of outdoor space used as educational resource. In fact, they used their gardens at schools, managed to convert abandoned spaces and developed their work in other surrounding green areas, therefore associating different types of outdoor spaces in the development of their project (Tables 2 and 3) (Tavares & Bettencourt, 2013).

Keeping in mind the participants' questionnaires, it was concluded that, before the course, there was a generalised lack of knowledge about the real potential of botanical gardens as privileged educational resources in science education, particularly concerning the loss of biodiversity and climate change topics. The attendance of the training courses and the development of projects enabled those teachers to increase knowledge of botanical gardens and other outdoor spaces, in addition to enriching and diversifying their set of educational tools, and increasing their level of confidence in the IBSE methodology and its application, based on a well-supported theoretical framework and varied types of activity.

According to them, the sharing of experiences and ideas among professionals, enabled them to acquire and/or consolidate knowledge and change their perception of the IBSE methodology. Therefore, the active learning methods used in preparing and teaching classes, the resources and materials, their relationship with students as well as students' results and their skills were improved, thus encouraging them to keep implementing IBSE.

These courses enabled them to reflect, improve and diversify their teaching practices and strategies and, the majority of the trainees, felt major differences after the application of the IBSE method. They reported that before, class preparation and the resources used were limited to explaining

Table 2. The Outdoor Spaces Used by Trainees of the Coinquire Course Editions.

COInquire – Outdoor Spaces used by Trainees	1st COInquire Course	2nd COInquire Course
Botanical gardens (BG)	7	4
School garden (SG)	14	10
Others	7	7
Total	28	21

Source: Tavares and Bettencourt (2013).

Table 3. The Outdoor Spaces Used Individually and Together By Trainees of the Coinquire Courses Editions.

COInquire – Outdoor Spaces used by Trainees	1st COInquire Course	2nd COInquire Course
Botanical gardens (BG)	2	1
School garden (SG)	7	5
Others (O)	3	4
SG + BG	4	3
SG + O	4	3
Total	20	16

Source: Tavares and Bettencourt (2013).

the contents of the textbook and restricted by the fulfilment of the pre-established curricular contents, resulting in a more formal and distant relationship between teacher and student.

After these courses they used more creative and diversified approaches, strategies and resources, and were able to involve other educational entities of the community (Table 4) (Silva et al., 2013).

Teachers stated they had a greater degree of confidence in their students' work, seeing that they now played a more active role in the classroom, therefore building a bidirectional and closer relationship between teacher and student in addition to enabling greater involvement in decision-making and collective work with positive impact on student outcomes (Table 4). Their students became more autonomous, involved, reflexive, critical, dynamic and able to become active actors in their own learning process (Silva et al., 2013; Tavares & Bettencourt, 2013; Tavares et al., 2013).

ASSESSMENT

Considering the overall results of the regular Educators course, participants rated this training as excellent. Knowledge acquisition about botanical gardens in different perspectives, in general, and their surroundings and the species that inhabit them, in particular, were the aspects that pleased the participants the most.

Despite this, the limited amount of time available to deepen some issues and the short duration of the course were less appreciated as participants wanted to develop the fieldwork further, in light of the importance of

Table 4. Evolution of the Environment in Class.

Questions	Before → Evolution	After
(A) Method used in the preparation and teaching of classes	– Textbook – Pre-established programme – Few practical activities – Poorly diversified strategies – Less freedom for students	– Variety of methodologies and learning strategies – More planning and critical reflection – More creativity – Active student involvement – Involvement of other subjects
(B) Resources and materials	– Textbook – Worksheets – PowerPoint – More teacher' documents – Pre-formatted resources – Less diversified resources	– Diversified resources – More practice – Involvement with other entities – Greater student involvement – More field work – More creativity
(C) Relation with students	– Formal and distant – Unidirectional (teacher–student) – Infantilisation of the student – More individualistic, less dynamic, less creative – Focused on the teacher	– Closer – Bidirectional (teacher–student and student–teacher) – Greater dynamism – More discussion and sharing – Better interpersonal relations
(D) Students' results	– Satisfactory or even good – More individualistic – Quantifiable objectives	– Better results – Greater argumentative ability – Better participation in class – More autonomy, power of observation and critical thinking – Improved behaviour in the classroom

Source: Silva et al. (2013).

showing these peculiar spaces, attracting the general public and strengthening their relationship with it.

From the results, the three-dimensional value of the explainers' role becomes clear: for themselves, for institutions and for their patrons. Also clear is the importance of these teaching–learning models and educational actions, which should be acknowledged and promoted on a regular and more specialised basis, at the level of higher education, not only due to its importance, but also because educators are the key links in the effectiveness of Communication and Science, as well as in Science Communication (Tavares, 2012; Tavares & Silva, 2014).

Taking into account the COInquire courses, participants showed a positive attitude towards IBSE and felt that their competence, participation, interest and motivation levels were enhanced.

In fact, when using IBSE methodology, both teacher and student individualities should be taken into account, respecting their uniqueness. This attitude is in line with the main characteristics needed to achieve a successful continuous professional development, as recommended in the INQUIRE project and courses, that is values and visions, mutual trust, respect, support and openness, collective responsibility for student learning, collaboration focused on learning (the group's as well as the individuals') (Bromley et al., 2013; Dillon et al., 2011; Kapelari et al., 2012).

With IBSE methodology, there is a positive relationship among students and between students and teachers, which is essential to develop a sense of personal responsibility as member of the community, promote a feeling of personal worth and compel students to become more inquisitive.

Science Education outdoors, using the IBSE methodology, revealed itself to be a 'natural' inquiry process in teaching–learning, as it focuses on working directly with living organisms (mainly plants), observing natural phenomena, formulating questions, linking evidence to explanations and finding appropriate solutions to explain observations and address questions and problems.

Globally speaking, after attending the COInquire course, teachers and educators, that is the participants, felt better about class preparation practices, methods and resources, and were able to reinforce their professional skills, as well as their relationship with students and their confidence in their students' work and results (Silva et al., 2013; Tavares & Bettencourt, 2013; Tavares et al., 2015).

REFLECTION AND FUTURE RECOMMENDATIONS

The outcomes and evaluation of the questionnaire indicates multiple benefits in studying Natural Sciences outdoors, resorting to the IBSE methodology.

This methodology, promoted by Inquire training courses for teachers and educators at the Botanical Garden of the University of Coimbra, is an innovative approach to curricular programmes on biodiversity, sustainability and climate change, and resulted in the improvement of students' performance and learning outcomes, revealing its potential to be real science education and professional development in practice, therefore confirming the importance of outdoor spaces for education.

As published in other papers (Silva et al., 2013; Tavares et al., 2014a; Tavares & Bettencourt, 2013), the Inquire IBSE training courses enhanced the trainees' professional competences, leading to replicable educational Natural Sciences case studies. Additionally, teachers gave their students more autonomy to create, experience, apply, make mistakes and repeat, and consequently gave them the needed freedom to build their own knowledge. The teacher–student relationship, based on mutual trust, was reinforced and consolidated with the application of this methodology. Using different strategies and designs, IBSE uses methodological pluralism, accepts difference instead of a single scientific design and learners keep experiencing the excitement of solving a question or problem on their own, or as part of a team (Tavares, 2013b).

However, teachers and educators must continue to update on the current scientific theories and facts about the different biology topics addressed, including any simple and basic educational actions, as is the case of the 'The algae that wanted to be a flower' lesson (Tavares, 2014).

Educational processes in outdoor live spaces, such as Botanical Gardens, parks, public, private or school gardens, encompasses direct interaction with living organisms, observation, question formulation, the connection to every-day life situations and solutions, in addition to 'naturally' following the IBSE methodology steps which are empowered when performed outside the classroom (Tavares, 2014; Tavares, Silva, & Bettencourt, 2015).

In fact, these spaces support active learning, that is an investigative process of knowledge acquisition, construction and understanding.

That is to say: the understanding of the learner regarding the world around them. Ask a question, find an answer, go further ahead, and improve. Asking a question means you want to know. The more you know the more there is to learn: that is the attitude of a scientist, using nature as the starting point and as everyday life practice.

If the stimuli are more challenging in an outdoor space and if an open and centred teaching learning methodology, as IBSE, is used then the two components – a living environment and a student-centred activity – are kept in mind in the research process which can mutually facilitate and improve the knowledge development process (Tavares, 2015a).

To fulfil and develop the main outdoor features of the IBSE method, as well as other techniques, trainers and trainees must be self-assured in their knowledge of these spaces to make the best use of these excellent educational resources to produce inventive IBSE projects about teaching natural sciences outside the classroom, as shown by the good results and advantages of the IBSE methodology used in the BGUC training courses (Tavares & Bettencourt, 2013).

Finally, based on this experience, the evidence described and the skills of the participants, arising from the IBSE courses outdoor, there is an incentive to develop an even more effective professional training, in the future, if the BGUC training course of educators, on resources and collections of the garden, could precede the IBSE Inquire courses.

Deepening knowledge and increasing educational practices in addition to exploring the possibilities of settings outside the classroom, as is the case of Botanical Gardens, allied with IBSE methodology will highly aid the development of Natural Sciences professionals' fieldwork and abilities, thus improving student knowledge acquisition (Tavares, 2015a; Tavares et al., 2015).

To ease an awakening to nature and its importance in world sustainability, let young children learn about wildlife in and with 'real' nature, outside the classroom, school or home. It is a challenge and a blessing to be able to accompany a child's first steps in their knowledge of the surrounding world and life. It is also a huge responsibility, as these footsteps mark the beginning of a life path and are pivotal in terms of the future scientific, civic, moral and political options and attitudes of each person.

Using IBSE methodology does not offer much to students, but provides them with an inspiring environment with the needed tools and opportunities to discover and use them on their own. Better than having answers, students ask questions. The less you give them, the more they will look for and find for themselves – and that is the only way to understand the links to knowledge, by ourselves – because then, the constructs and products of our efforts and work are already there, within our reach. Provide them with motivation – the thrust to go further ahead, experience things and acquire new knowledge – which is the switch that feeds the mind. Adults, teachers and educators were once children, and this gives them the ability to understand and follow their learners, respecting and dazzling them with different and diverse identities and realities.

While monitoring their students' inquiry, analysis, inference and argument (Tomkins & Tunnicliffe, 2007), a teacher or educator must be attentive, aware and prepared to follow and also learn with their students, evolving and maturing their own background and possibly their wisdom – the best educational tool –, a kind of understanding refined by practice.

ACKNOWLEDGEMENTS

To the students of the University of Coimbra, and other similar institutions, who accompanied for almost two decades and attended Botany and Education on Science in the BGUC, reaching all sorts of public, especially schools.

To the INQUIRE Educative European Project manager board, partners and Coimbra team, for the great educational and innovative experience, extending Natural Science learning and teaching outdoors through IBSE methodology.

The linguistic correction by a native-speaker supported by FCT/MEC through national funds (PIDDAC) and co-financed by FEDER through the COMPETE – Operational Programme Competitiveness Factors under the project PEst-C/CED /UI0194/2013.

REFERENCES

ASE OSWG. (2011). Association for Science Education. Outdoor Science Working Group. Outdoor Science. Shrewsbury: Field Studies Council and King's College of London. Retrieved from http://www.nuffieldfoundation.org/sites/default/files/files/ase-outdoor-s

Bora, S., McCrea, E., Gay, M., Herrmann, L., Hutchison, L., Pistillo, M., & Wirth, S. (2010). *Early childhood environment education programs: Guidelines for excellence*. Washington, DC: The North American Association for Environmental Education Publications.

Bromley, G., Regan, E., Kapelari, S., Dillon, J., Vergou, A., Willison, J. … Tavares, A. C. (2013). *O manual do curso inquire*. Retrieved from http://www.inquirebotany.org/uploads/files/FCTUC%20Portugal/FCTUC%20-%20manual%20do%20curso%20INQUIRE%20COIMBRA.pdf

Cachapuz, A., Praia, J., & Jorge, M. (2002). *Ciência, educação em ciência e ensino das ciências*. Lisboa: Ministério da Educação.

CIEEP (Coimbra Inquire Educative European Project). (2014). Retrieved from http://www.uc.pt/jardimbotanico/projetos/inquire/

Dillon, D., Elster, D., & Kapelari, S. (2011). Document summarizing how IBSE is defined in the INQUIRE course. INQUIRE EU 7th Framework Deliverable, London, UK.

Duit, R., & Treagust, D. F. (2003). Conceptual change: A powerful framework for improving science teaching and learning. *International Journal of Science Education*, *25*(6), 671–688.

Exploratorium. (2014). *Field trip explainer program*. Retrieved from http://www.exploratorium.edu/education/field-trip-explainers

Gatt, S., Tunnicliffe, S., Borg, K., & Lautier, K. (2007). Young maltese children's ideas about plants. *Journal of Biological Education*, *41*(3), 117–121.

Inquire Educative European Project [IEEP]. (2013). Retrieved from http://www.inquirebotany.org/

Kapelari, S., Bonomi, C., Dillon, J., Regan, E., Bromley, G., Vergou, A., & Willison, J. (2012). The INQUIRE train the trainers course manual. London, UK.

Kapelari, S., Vergou, A., Willison, J., Regan, E., & Dillon, J. (2014). Strengthening networks for people, plants and botanic gardens. In N. Krigas, G. Tsoktouridis, C. M. Cook, P. Mylona, & E. Maloupa (Eds.), *European botanic gardens in a changing world: Insights into Eurogarden* (Vol. 6, pp. 273–279). London: Balkan Botanic Garden of Kroussia and BGCI.

Linn, M. C., Davis, E. A., & Bell, P. (2004). *Internet environments for science education*. Mahwah, NJ: Lawrence Erlbaum Associates.

Lopes, L., & Bettencourt, T. (2012). Inquiry based science education: Uma visão global. In *Livro de resumos do 1º Fórum Nacional COInquire*, (p. 3). Retrieved from http://sequoia.bot.uc.pt/jardim/inquire/forum2012/files/111forum_inquire_21012_livroresumos.pdf

National Research Council. (2009). *Learning science in informal environments: People, places, and pursuits – Free executive summary*. Retrieved from http://www.upf.edu/pcstacademy/_docs/InformalLearningReport.pdf

Page, L. M. (2005). Organisms in nature as a central focus for biology. *Trends in Ecology and Evolution*, *20*(7), 361–362.

Pathway Educative European Project. (2011). Retrieved from http://www.pathwayuk.org.uk/what-is-ibse.html

Patrick, P., Byrne, J., Tunnicliffe, S., Asunta, T., Carvalho, G., Havu-Nuutinen, S., & Tracan, R. (2013). Students (ages 6, 10, and 15 years) in six countries knowledge of animals. *Nordina*, *9*(1), 18–32.

Patrick, P., & Tunnicliffe, S. D. (2011). What plants and animals do early childhood and primary students' name? Where do they see them? *Journal of Science Education and Technology*, *20*(5), 630–642.

Pessoa, P. (2013). *Ensaios de voos-metamorfoses pessoais e profissionais*. Porto: Tese de mestrado não publicada. Universidade Católica Portuguesa.

Rudd, T. (2008). *Reimagining outdoor learning spaces*. Bristol: Futurelab.

Science Museum. (2014). Retrieved from http://www.sciencemuseum.org.uk/educators.aspx

Scientix European Plataform. (2014). Retrieved from http://www.scientix.eu/web/guest/projects/project

Silva, S., Tavares, A. C., & Bettencourt, T. (2013). The IBSE methodology at Botanic Garden of Coimbra – results and reflections about two courses undertaken. In *Proceedings of the 6th international conference of education, research and innovation* (ICERI 2013). Retrieved from http://library.iated.org/view/SILVA2013IBS

Tavares, A. C. (2008). BotaniCOI: A calendar of educative activities in the garden. *O/El Botânico – Revista da Associação Ibero-Macaronésica de Jardins Botânicos*, *2*, 28–30. Retrieved from http://www.elbotanico.org/revista2.html

Tavares, A. C. (2012). Sementes de jardim: disseminação de cultura científica. In *Livro de resumos do 1º Fórum Nacional COInquire*, (p. 29). Retrieved from http://sequoia.bot.uc.pt/jardim/inquire/forum2012/files/111forum_inquire_21012_livroresumos.pdf

Tavares, A. C. (2013a). *A alga que queria ser flor/The algae who wanted to be a flower.* Coimbra: Imprensa da Universidade de Coimbra.

Tavares, A. C. (2013b). Explorers in the botanic garden: An IBSE model on biodiversity and climate change. *O/El Botânico – Revista da Associação Ibero-Macaronésica de Jardins Botânicos, 7*, 47–50. Retrieved from http://www.elbotanico.org/revista7.html

Tavares, A. C. (2014). The Alga who wanted to be a flower – An outdoors IBSE model on plants evolution. Hands-on science. Science education with and for society. In M. F. M. Costa, P. Pombo, & B. V. Dorrío (Eds.), *Hands-on science network* (pp. 19–28). Minho, Aveiro and Vigo: Universidade do Minho, Universidade de Aveiro and Universidade de Vigo. Retrieved from http://www.hsci.info/Book_HSCI_2014_lowresA4.pdf

Tavares, A. C. (2015a). *Educação em Jardins Botânicos: 16 anos de experiência.* Málaga: Eumed – Universidade de Málaga. Retrieved from http://www.eumed.net/libros-gratis/2015/1442/index.htm

Tavares, A. C. (2015b). *UM TRIBUTO AO JARDIM BOTÂNICO DE COIMBRA. Património Mundial da Humanidade, 2013/A TRIBUTE TO THE BOTANIC GARDEN OF COIMBRA. World Heritage Site, 2013.* Málaga: Eumed – Universidade de Málaga. Retrieved from http://www.eumed.net/libros-gratis/2015/1465/index.htm

Tavares, A. C., & Bettencourt, T. (2013). Inquire trainer´s courses: a pool of IBSE projects on teaching natural science outdoors. In *Proceedings of the 6th International Conference of Education, Research and Innovation* (ICERI 2013). Retrieved from http://library.iated.org/view/TAVARES2013INQ

Tavares, A. C., & Silva, S. (2014). The valuable role of educators. *O/El Botânico – Revista da Associação Ibero-Macaronésica de Jardins Botânicos, 8*, 71–73. Retrieved from http://www.elbotanico.org/revista8.html

Tavares, A. C., Silva, S. & Bettencourt, T. (2013). Aftermath of the two editions of INQUIRE training course on IBSE methodology at Coimbra Botanic Garden. In *Proceedings of Inquire Conference Raising Standards Through Inquiry: Professional Development in the Natural Environment* (pp. 10–14). Retrieved from http://www.inquirebotany.org/en/resources/inquire-conference-proceedings-256.html

Tavares, A. C., Silva, S., & Bettencourt, T. (2015). Advantages of applying IBSE method: the Coimbra Inquire Course case-study. 6th WORLD CONFERENCE on Educational Sciences (WCES 2014), Procedia-Social and Behavioral Sciences, 191, 174–178 (Elsevier ISSN: 1877-0428).

Tavares, A. C., Silva, S., Santos, J., Paiva, I., Oliveira, J., & Bettencourt, T. (2014a). Inquire at Coimbra Botanic Garden: Products and Process of an IBSE Educative Project. *Procedia-Social and Behavioral Sciences, 116*, 4353–4356.

Tavares, A. C., Silva, S., Santos, J., Paiva, I., Oliveira, J., Figueiredo, A., & Bettencourt, T. (2014b). Inquire: Why teaching about the conservation of biodiversity and climate change? In N. Krigas, G. Tsoktouridis, C. M. Cook, P. Mylona, & E. Maloupa (Eds.), *European Botanic Gardens in a Changing World: Insights into Eurogarden* (Vol. 6, pp. 281–290). London: Balkan Botanic Garden of Kroussia and BGCI.

Tomkins, S., & Tunnicliffe, S. D. (2007). Nature tables: Stimulating children's interest in natural objects. *Journal of Biological Education, 41*(4), 150–155.

Tunnicliffe, S. D., & Ueckert, C. (2007). Teaching biology – the great dilemma. *Journal of Biological Education, 41*(2), 51–52.

Wenning, C. J. (2005). Implementing inquiry-based instruction in the science classroom: A new model for solving the improvement-of-practice problem. *Journal of Physics – Teacher Education Online, 2*(4), 9–15.

INQUIRY-BASED LEARNING IN ACTION: REFLECTIONS ON AN INTERDISCIPLINARY MASTER'S PROGRAM IN THE HEALTH AND LIFE SCIENCES

Marjolein B. M. Zweekhorst, Wanda S. Konijn, Jacqueline Broerse and Jeroen Maas

ABSTRACT

Society is increasingly confronted with a range of complex social problems that need to be addressed using a research process based on collaboration between stakeholders from both science and society and the integration of knowledge from different disciplines. This type of interdisciplinary research is more complex than mono disciplinary research and requires skills at the cognitive, inter-personal, and intra-personal levels. We present the experiences with an interdisciplinary master's program. The research question we address is what educational strategy prepares students for interdisciplinary research on complex social problems? Since tasks which are too complex can frustrate students and create resistance,

Inquiry-Based Learning for Science, Technology, Engineering, and Math (STEM) Programs: A Conceptual and Practical Resource for Educators
Innovations in Higher Education Teaching and Learning, Volume 4, 171–191

ISSN: 2055-3641/doi:10.1108/S2055-364120150000004011

we argue for a gradual approach to inquiry-based learning. We interviewed both students and lecturers, and included curricula evaluations. We found that students can be trained in interdisciplinary research based on a gradual approach to open inquiry and we found a relationship between the complexity of cognitive tasks and the amount of learning in other domains. We argue that when students are challenged at the right level with appropriate guidance, the learning domains will reinforce each other. To keep students optimally challenged, it is crucial that the teachers adjust their role while directing students from structured inquiry towards open inquiry.

INTRODUCTION

Society is increasingly confronted with a range of problems, which impact on health and well-being: combatting HIV/AIDS, the vicious circle of poverty and poor health, and comorbid conditions in an aging population. These problems are complex and, consequently, difficult to deal with since they require the actions of different stakeholders with different interests, views, and needs (Klein et al., 2001). As these types of complex problems do not fit into specific scientific disciplines, they are complicated to study. A research process is therefore needed for these kinds of problems which is capable of integrating knowledge from different disciplines and of supporting collaboration between stakeholders from both science and society (see e.g., Bunders et al., 2010; Klein et al., 2001). Researchers who are capable of conducting interdisciplinary research are in increasing demand.

To study a problem from different disciplinary perspectives requires certain cognitive skills; students need to be able to integrate social and scientific theoretical models. In addition, the problem studied may be the object of intense debate and the target of strong societal forces, like political parties, media, and interest groups. Students, who will become interdisciplinary researchers, therefore, require an understanding of the social dynamics and basic psychological processes in organizations and society and organizational change. They also need to be able to deal with different personalities and cultures, and to adjust their strategy as required. Therefore, interdisciplinary research not only forms a challenge at the cognitive level, but also on the inter- and intra-personal level (for example explaining the value of social scientific models to scientific researchers and lay interest groups). However, little is known on how to prepare students for interdisciplinary research on complex societal problems.

In this chapter we describe the lessons learned from eight years of students' and teachers' experiences with the two-year interdisciplinary master's program "Management Policy Analysis and Entrepreneurship in the Health and Life Sciences" offered by the VU University, Amsterdam, the Netherlands. The program aims to prepare graduates to conduct research at the nexus of science and society, and to contribute to solving complex societal issues related to the health and life sciences. The Management Policy Analysis and Entrepreneurship in the Health and Life Sciences program applies an inquiry-based approach of "problem-driven education", which focuses on complex problems in the health and life sciences throughout the program. We see students as active learners who are responsible for the development of their own knowledge (e.g., Tynjälä, 1999). The research questions we address in this chapter is to what degree does our inquiry-based educational strategy prepare students for interdisciplinary research on complex social problems and does it succeed in training them at the cognitive, inter- and intra-personal levels at the same time?

We start by explaining the educational approach applied within this program. Next, we describe the methodology used in our study. We discuss both the difficulties and the benefits of learning experienced by students in the cognitive, inter-personal, and intra-personal domains and indicate how learning within each domain influences learning experiences in the others. We describe the challenges of the educators with the inquiry-based master's program and discuss strategies to improve the learning at the three levels. We focus on lessons learned on how to prepare students for cross-disciplinary, open inquiry in interdisciplinary research.

BUILDING AN INQUIRY DRIVEN INTERDISCIPLINARY MASTER'S PROGRAM

Inquiry-based approaches seem to be the most suitable to prepare students for addressing complex problems. From the 1960s onwards, instructional approaches have shifted from more traditional forms of teaching (memorizing and reproducing) to more self-directed approaches to foster learning (see, e.g., Bruner, 1961; Dewey, 1933; Piaget, 1964; Vygotsky, 1978). Various methodologies such as Problem Based Learning (Barrows & Tamblyn, 1980; Savery, 2006), experiential learning (Kolb & Fry, 1975), and Project Based Learning (Barron & Darling-Hammond, 2008) were developed as a result. Most of these approaches start with a problem and the guiding idea of all these approaches is to engage the student in a process of "learning by doing." Students are, in most cases, faced by situations in which they have to discover and construct knowledge through their own activities (Prince & Felder, 2006; Spronken-Smith, 2007). An overview of 33 learning approaches that start with a problem is given by Woods (2014).

The various inquiry-based approaches differ in the amount of guidance offered by the instructors, with unguided discovery learning at one end of the spectrum and Process-Oriented Guided Inquiry Learning (Farrell, Moog, & Spencer, 1999) at the other. Different researchers support or critique either end of the spectrum. Kirschner, Sweller, and Clark (2006) argue that, if students do not have sufficient prior knowledge, it is important that guidance is provided to support the cognitive process of learning. We take this approach with our students and also argue for a more guided approach to inquiry. In the Management Policy Analysis and Entrepreneurship in the Health and Life Sciences program we apply a teaching strategy which is based on the four stages of inquiry described by Banchi and Bell (2008).

Banchi and Bell (2008) promote inquiry-based learning. However, they argue that it is not possible for elementary students to conduct a scientific investigation from the start, since this is too complex. They argue that students need more practice to be able to do that. Banchi and Bell (2008) recognize that there are various levels of inquiry and that, by taking students through these various steps, the students can be prepared for scientific investigations. They classify a continuum of four levels for inquiry in a certain activity: confirmation, structured, guided, and open. Every level has a particular amount of information provided to the students and a certain level of guidance given by the teacher.

The first level is confirmation inquiry. At this level a lot of guidance is provided by the teacher, including the research question and the proposed procedure. Banchi and Bell (2008, p. 26) mention three reasons why this inquiry level is useful: "a goal is to reinforce a previously introduced idea; to introduce students to the experience of conducting scientific investigations, or to have students practice a specific inquiry skill ...." At the next level, structured inquiry, the question and procedure are still provided by the teacher, but now the students have to generate data and explain how they collected them. Banchi and Bell (2008) argue that students need to master these lower levels of inquiry first in order to develop the skills required for higher levels of guided and open inquiry. At the level of guided inquiry the teacher only provides the research question and the students have to design the procedure to test their question and the resulting explanation. The guidance of the teacher is still crucial. At the final level of open inquiry the students act like scientists. They develop the questions, design and carry out the investigation and communicate their results relatively independently. We applied the four stages of inquiry, described by Banchi and Bell for teaching for elementary students, to teaching how to address complex social problems within the interdisciplinary master's program Management Policy Analysis and Entrepreneurship in the Health and Life Sciences.

The Master's Program Management Policy Analysis and Entrepreneurship in the Health and Life Sciences

The Management Policy Analysis and Entrepreneurship in the Health and Life Sciences program aims to prepare students for conducting interdisciplinary research and thus needs to take the three domains – cognitive, inter- and intra-personal skills – into account. The program provides a broadening of the knowledge and skills from a bachelor of science degree (such as biology, medical biology, health sciences, or medicine) to disciplines such as, science, technology and society studies, policy science, and management studies. The students need to use interdisciplinary research methodologies as well as to contrast and integrate scientific and social scientific paradigms with expert knowledge. Teamwork, in which both inter- and intra-personal skills can be developed, is an important aspect of the training. The first year of the master's program aims to teach the following core competences: analysis of complex societal issues; formulation and implementation of strategies to deal with complex societal problems by way of interdisciplinary research; effective cooperation and communication with researchers from scientific disciplines other than health and life sciences and with societal actors. The year consists of three compulsory courses of four weeks or six European Credits (EC, which comprise 168 hours of study), 12 EC elective courses, and a research internship (30 EC). The aim of the second year is to deepen the students' research skills within one of the four specializations (Policy, Management and Entrepreneurship, International Public Health and Communication all in the Health and Life sciences) and to prepare them for conducting research independently. This chapter focuses on the first year of the program in which the students are brought from level 1 to 4, which is described in more detail now.

Confirmation Inquiry Course

The Management Policy Analysis and Entrepreneurship in the Health and Life Sciences program starts with the course "Research methods for analysis of complex social problems." The aim of this course is to introduce students to conducting investigations within the interdisciplinary research field. We therefore apply the first level of inquiry, confirmation inquiry. We start with training at the cognitive level. The course starts with a complex societal challenge such as: "How to improve the health of the poor in Ethiopia?" Students realize that this question cannot be answered within one scientific discipline and students assess the various perspectives that different scientific disciplines would offer. The lecturer presents a range of theoretical models and research paradigms, which can offer simplified representations of the reality. In addition, students are trained both in qualitative and quantitative data collection methods, which they also practice in role plays (e.g., interviews). Students work in small teams of four people, with different bachelor backgrounds, and apply these frameworks to the complex research questions. At this stage, little explicit attention is paid to the inter- and intra-personal skills.

Structured Inquiry Course

In the second course, we make the transition to structured inquiry. In the course "Analysis of Governmental Policy," students are commissioned by an external organization to work on a complex societal problem related to the health and life sciences, which is facing the organization in question. To help students gain a basic understanding of this new scientific field, policy science, the students watch short videos on a variety of health policy topics. In a series of masterclasses the students work on various assignments to engage with the topics from the videos in order to get a better understanding of them. As a team, the students need to conduct interdisciplinary research in order write a policy advisory report for the commissioning party on the basis of the literature and interviews with various stakeholders. An example of such an assignment issued by the Ministry of Health could be: "Should the Ministry include the vaccine for pertussis (whooping cough) in the national vaccination program?" Explicit attention is paid to both cognitive and inter-personal skills (team work and face-to-face interviews).

In teams of ten, students have to select a theoretical framework by themselves and prepare a design and plan for their research. The team facilitator provides feedback and, if necessary, the students adapt the research design. Next, the team has to divide the work and collect information, via literature and interviews with various stakeholders. The students find that different stakeholders from different disciplines and professions give different answers to the same questions. The lecturer pays specific attention to the integration of science-specific knowledge with the knowledge and skills of other disciplines of the social sciences and to teamwork. The team has to integrate the information and reach consensus on the policy advice. Finally, students need to set up and to translate research findings into policy recommendations with valid lines of argument. At the end of the course, students present the policy advice to the representative of the external institution.

Guided Inquiry Course

The third course, Communication, Organization, and Management, aims to provide students with a deeper understanding of the management process of translating scientific knowledge to complex societal problems while using a more open-ended guided inquiry approach. Attention is paid to both cognitive skills and intra-personal skills. Students are introduced to some trends in today's changing landscape of health care. Telemedicine, financial changes, new technologies, changing populations, re-emerging diseases are just a few of these trends that are bound to transform the way healthcare systems function. In order to quickly obtain some background in organizational studies, the teacher provides an overview of the most prominent concepts and theories in this field as a starting point for the students. In working groups of about twenty students, a facilitator helps the students explore the value and applicability of organizational theories. In groups of five, the students then receive an assignment, which is less guided than the assignment of the previous course, to prepare them for their internships. In this assignment the students act as managers in a hospital and need to decide which trends are likely to be the most influential for their hospital and what the foreseeable consequences may be. They need to assimilate more facts and go into organizational theories in more depth. Based on that assessment, they need to specify their research question and determine what actions management needs to take in order to address these future consequences. Finally, they need to present their assessment in an argument that persuades their fellow managers of the hospital to apply the changes they deem necessary. Students use peer feedback to improve their documents. This is the final course which prepares the students for the uncertainty that interdisciplinarity brings.

Elective Course and the Internship Research (Open Inquiry)
After the three compulsory courses students choose two elective courses of six EC each. In these courses they apply guided inquiry in their field of specialization. Each elective involves some training in additional research methods. After the two elective courses the students are engaged in a process of *open inquiry*, during their five months individual internship research on a complex societal problem. Students may search for a research project themselves or apply for one offered through the internship vacancy database of the department hosting the MPA program. The examination board of the program assesses whether the internship fulfills the requirements of the program. The research starts with writing a research proposal. Every week the lecturer and the student meet to discuss progress. Students have six weeks to write the detailed design for their research. After six weeks a critical assessment takes place to assess whether or not to continue with the internship or to extend the time needed to write the design (up to a maximum of two weeks, in order to allow enough time for data collection). If the research is to be continued, students now need to demonstrate their capacity in all three domains. They start collecting data and conduct interviews by themselves. Next, students compare and relate their data with data from other sources and the literature, they discuss their findings and draw conclusions. During the presentation of their research they not only have to summarize and describe their findings and interpretations, but also have to defend and justify their conclusions. Students are assessed on their final report, presentation, and academic attitude and professional behavior. At the end of the internship students reflect on their internship with respect to all three levels of learning. Throughout the research the lecturer is involved as a coach and advisor.

METHODOLOGY

In-depth interviews were undertaken by the second author with 26 selected students of the 2013–2014 cohort in groups of two at the end of the second compulsory course. In the course there were 13 groups of 10 students. We asked a colleague, who has not been involved with the students before, to select two students from each group on the list we provided. These 26 selected students received an invitation by e-mail. We also interviewed six students of the same cohort after the third course. We asked a colleague, who has not been involved with the students before, to select six working groups. In these working groups we asked for voluntary cooperation in an interview; for each session one student volunteered. Interviews focused on various aspects; aims of the course, new insights they acquired, benefits from the course, the difficulties encountered and the differences from previous learning experiences.

In addition, in-depth interviews were undertaken with several lecturers of the program, including those responsible for the three compulsory courses. These interviews focused on the lecturers' perspectives on the highlights of the courses, the difficulties the students experienced on the cognitive, inter- and intra-personal levels, and the challenges of teaching more open inquiry. All interviews were transcribed by the second author and analyzed by the first two authors (see for the interview topic list appendix).

We also used the yearly curricula evaluations. During these evaluations, students evaluate the curriculum with various qualitative evaluation methods; such as photovoice (Berg, 2004; Wang & Burris, 1997) and group discussions. In order to assess whether the educational strategy succeeded in producing in interdisciplinary researchers we also include the result of the external program evaluation executed by the Dutch-Flemish accreditation committee.

RESULTS

Confirmation Inquiry Course

The main teacher of this course emphasized that this is a very important course for making the step towards more interdisciplinary thinking. This can also be seen in students' responses. In curricula evaluations at the end of the program, students stated that the first course was very valuable as it enabled them to conduct interdisciplinary research and to understand the value of the frameworks. They saw the value of interdisciplinary teams and how team members could learn from each other due to their different backgrounds.

The lecturer mentioned that students in this course face difficulties with understanding the various theoretical frameworks. All students have their previous training in the natural and life sciences in which they normally apply different singular scientific paradigms. They are now faced with the option of choosing a research framework from a variety of disciplines, all of which may be valid. In interviews, the lecturers reported that for most students this is new and difficult to understand. This can also be seen from the interviews with students. Interviews at the end of the second course show that they had been confused during the first course, as illustrated by the following quote: "Yes, I am … a chemistry student and that is almost exclusively beta [sciences] oriented and then you come here and get introduced with so many things [various conceptual frameworks] from social sciences: that takes some getting used to" (TG 10 R1).

One student remembered that he thought he understood the application of the theoretical frameworks after the first course, but after the second course realized that he did not. The second course resulted in a deeper understanding, as is illustrated by the following quote: "Now after the [second] course I actually have the impression that I really understand [applying theoretical frameworks]" (TG 3 R1).

The lecturer said that it was important to provide a lot of structure on how to conduct the assignment. In the assignment the students have to apply theoretical frameworks to a complex research question. The lecturer presents and explains to the students the various possible frameworks in relation to the complex question so that they can realize the advantages and disadvantages of each (confirmation inquiry).

Training students at a cognitive level influenced the two other domains of learning during the first course. At the cognitive level, students were facing new models and paradigms. The fact that they were suddenly faced with multiple options – which could even be adapted – was new. For some students, the novelty and multiplicity of these paradigms created a complex working environment in which they felt insecure. This affected learning in the intra-personal domain, since they had to deal with uncertainty. The level of uncertainty experienced by the students varies depending on their background. Students with a bachelor degree in for example Liberal Arts and Sciences were more familiar with interdisciplinary research, and thus less insecure. The students appreciated and valued the differing bachelor backgrounds of their fellow students.

Structured Inquiry Course

Lecturers reported that the students became very enthusiastic during the second course, especially after meeting the commissioners of the research. Students said that working on a real-life project inspired them. They also said that being given a task by a commissioning party was highly motivational. In addition, they appreciated the teamwork. They also more appreciated the benefit of the first course, in which they gained insight into the frameworks of various disciplines: "It is one of the good things in the research, that together [with other students] you learn to work with other perspectives in the field. It is real contact with people" (R1). In the curriculum evaluation at the end of the year, students said that the learning at the inter- and intra-personal levels had been very valuable.

Various difficulties were, however, faced by the students during the course. Students were faced with an additional range of models from the policy sciences. Lecturers reported that students questioned why they needed to know all the models and regularly asked the lecturer whether they had made "the right" choice; students felt insecure about the choices they made. Next, conducting interdisciplinary research was difficult for the students. In interviews students were faced with different stakeholders. This reminded the students of the personalities they enacted in the role-plays during the first course (the person that does not like to speak, the person that keeps on talking etc.), but students said that this did not always stop them from repeating the mistakes they had made in the role-plays. In addition, integrating the various perspectives encountered into a single report was experienced as a challenge by most of the students.

In this course the students are provided with some structure (structured inquiry). Students are provided with the research question, the methodology, and the context. However, they have to conduct the work themselves. The lecturers reported that in the beginning of the course, the students did not see the benefit of working in a group and the need for daily discussions amongst their group members. They preferred to conduct the assignment alone. An explanation of the amount of work, which can be done with 10 people in one month (similar to one internship of one person for five months) helped. However, working with 10 people requires regular discussion on the progress and findings. Therefore, with help from group coaches, time was spent on the functioning of the team. Students assessed their role within the team and how to improve the effectiveness of the team. The difficulty faced by the group coaches was to provide enough structure for the students, but to leave enough room for the students' own input. Lecturers also mentioned the complexities of the assessment of inter-personal skills, such as demonstration of leadership, active communication etc.

In this course, students learned a lot in the inter-personal domain. They realized they needed each other to jointly develop the policy report. Effective communication between the team members was important. In addition, good communication and listening skills were necessary for collecting the information from the various stakeholders. This directly influenced learning at the intra-personal level. By working in a team, all the students had to adapt to the team. Some students were not comfortable expressing their views within the team, however, it was essential that they did so as they had collected part of the information. More confident students, on the other hand, had to take a less dominant position. The interviews with other stakeholders stimulated students to critically assess their own role in the process, which stimulated learning at the intra-personal level. Learning at the inter- and intra-personal levels in turn stimulated learning at the cognitive level by deepening the understanding of the research tools and the application of the framework.

Guided Inquiry Course

The lecturer mentioned that students preferred the freedom in this course. Most students confirmed this by stating that they liked the freedom to work in a more creative way and also appreciated the training in feedback. The students provided feedback to one another regarding their assignment and the way they cooperate. Attention was paid to formulation and content of the feedback and to how students reacted to it. This is an important contribution to interdisciplinary research and group work. Students realized: "That you learn to give others clear feedback without hurting, or insulting, them or something else in any other way .... Or you have the feeling that you miss something and then you become insecure ..." (IW2). And another quote: "Hmm, I analyze directly what happened in the previous course. This was, that was really a form of group thinking, in vulnerability and the exclusion of new ideas, those kind of things and then I felt, really a little down actually. Yes, thus, but I found it useful ..." (I W 2).

According to both the interviewed lecturer and students, the difficulty of the course is that the freedom and the flexibility the students have in designing their assignment increases complexity and makes some students feel insecure about whether they are taking "the right" decisions. The assignment is chosen to resemble a real-life complex problem. The students then have to make many choices with respect to the assignment, on which trend or on which component of the organizational structure they want to focus. The struggle some students have with the assignment during the beginning of the course is addressed in different ways. Some students just start analyzing and writing, while others seek repeated assurances (by taking students from previous years as peer examples) that they can do it as well. The coaches play a role in reassuring the students that they have the capacity to conduct the assignment, further enhancing their learning.

The assignment in this course has less structure than the previous course (structured inquiry). The lecturer said that he learnt over the years that it is important to help the students get on with the intra-personal learning element of dealing with uncertainty. The lecturer explains to the students that it is a process like organizational change. Some accept it from the beginning whereas others have to adapt and go through a process much like that described in common change curves (e.g., Elrod & Tippett, 2002), including phases of denial and bargaining (when they do not know what to choose), and acceptance and creativity (when they see their choices lead to solutions). During the course the lecturer reminds the students of these phases and in the lectures there is time for discussion to see how they feel – and where they position themselves on the change curve. According to the lecturer this is very helpful to the process. This lecturer also mentions the challenges with the assessment of the inter- and intra-personal skills.

In this course students immediately use the cognitive knowledge they gained, and apply it in the inter-personal domain. They have to deal with feelings such as denial, they have to bargain and negotiate. Further intra-personal learning takes place during the discussions in the working groups and through giving and receiving peer feedback. Tolerance for stress and ambiguity as well as a critical assessment of the students' own role are required.

The Optional Courses and the Internship (Open Inquiry)

Most students are enthusiastic about the optional courses and the internship. Through the optional courses the skills and competences in interdisciplinary research of the students are strengthened. Lecturers indicate that if students successfully pass the compulsory courses, they are able to write a research design for their internship and conduct the internship research successfully.

Selection of a suitable framework for their internship research remains, for most students, a real challenge. Although most students are able to collect their data from various sources (literature, interviews with various stakeholders, etc.), integration of the various sources of information is still challenging for some. The research outcomes also differ in quality. Some students are able to write beautiful reports (and even scientific articles), while some need a lot of assistance and have to work very hard to get a pass.

We found that during the first six weeks it was important to provide some structure by refresher lectures and weekly meetings in which the students discussed their progress and challenges faced with fellow students and their supervisor. Many students said that it is very helpful that other students face similar problems at the same time and therefore that they can support each other. Most lecturers said that it is important to find a balance between providing structure and remaining flexible and that for some students open inquiry is still complex.

During the internship students have to work in a complex social environment. Most students are on the one hand excited to conduct their internship research, but also a bit anxious to really start. In the end all the students learn within all domains, but for every student and for every internship the precise learning within these domains is different.

Curricula Evaluations

The yearly curriculum evaluations showed that students are proud of their progress during the first year. Although they mention learning in the cognitive domain, they especially emphasize the learning in the inter- and intrapersonal domains.

In 2012 the master's program was reaccredited by the Dutch-Flemish accreditation committee, an international external accreditation committee (NVAO, 2012). Overall, they were very positive about the program. They assessed it on the basis of three standards concerning: Education learning environment; Realized end qualifications; and Examination and final level; with, respectively, good, good and excellent. The committee made some explicit remarks concerning the learning environment: "The concept is problem-oriented education. The didactical concept is being applied strictly in order to stimulate the learning capacity and creativity of the student" (NVAO, 2012, p. 16) and "The students reach, as judged by the panel, the final qualifications during the program. The coherence and the structure of the program with the increased complexity of the courses is positive. The relation between the courses and the internship is very powerful" (NVAO, 2012, p. 16).

DISCUSSION AND CONCLUSION

The main questions addressed in this chapter were, to what degree does our inquiry based educational strategy prepare students for interdisciplinary research on complex social problems and does it train them at the cognitive, inter- and intra-personal levels at the same time? In this chapter we described how an inquiry-based educational strategy prepares students for interdisciplinary research on complex social problems. We applied the four stages of inquiry, described by Banchi and Bell (2008) for science teaching for elementary students, to teaching how to address complex social problems within an interdisciplinary master's program. For this type of interdisciplinary research it is important to train students not only in cognitive skills, but also inter- and intra-personal skills. We found that inter- and intra-personal skills can be classified and learnt according the four levels identified by Banchi and Bell (2008).

Cognitive

Coming from a beta science background, most of the students have been taught in a specific scientific paradigm, in which it is clearly defined what direction research should take and how it should be performed. Being faced with multiple frameworks from different disciplines, which all provide different answers, is for some students difficult to understand and can make them feel insecure. Therefore, we start the program with confirmation inquiry aiming to introduce interdisciplinary research and theories and models from a variety of disciplines. In the second course, we apply a structured approach to inquiry where students learn to collect, interpret, and integrate data from various sources in order to answer the research question provided by the teacher. In the third course, we apply a guided inquiry learning approach. Students have to specify the research question and integrate knowledge in order write a convincing report. During the internship the students have to conduct all the research steps: definition of research question, selection of the framework, data collection, analysis and comparison with the wider literature, as well as the discussion and conclusion.

Inter-Personal

During the first course students learn to value and appreciate different bachelor backgrounds of other students. In the next course, the group size is increased to 10 students, which makes it much more difficult to work together. The structured approach allows the students to practice the cognitive skills learnt during the first course, and work on the inter-personal learning tasks of working in a team. Students have to practice their listening skills and to work at effective communication both within the team and with external stakeholders. In the third course, in which they are only guided in the approach and still work in a team, skills such as conflict management and negotiation become more dominant in the program. In addition students learn to give (peer) feedback. In the assignment students have to write a convincing report to convince the management of the benefit of their plan, which forces them to substantiate their arguments and explain their choices. During the internships, students have to collaborate with a diverse set of stakeholders.

Intra-Personal Skills

In the first course students question why their beta bachelor program is not sufficient. Why do they have to learn all these new frameworks and why can't the teacher just provide the correct model? In this course we apply confirmation inquiry, and the students need to overcome a lot of hesitation in order to become motivated. Starting from the second course, structured inquiry is applied. To work effectively in the team requires critical self-reflection and adaptation. Moreover, the interviews require a critical reflection on the student's own role and attitude. In the third course we apply guided inquiry. In this course the intra-personal skills are stimulated further by a less structured assignment. Students provide peer feedback to their fellow students, which requires good interpersonal skills. Finally, students have to conduct their internship in open inquiry. Most students need to be encouraged to explore their insecurity and find a way to overcome it. We have categorized the learning in the three domains at the four levels of inquiry in table one, which can also serve as the further development of a taxonomy for including inter- and intra-personal learning.

To educate interdisciplinary researchers we argue for a gradual progress towards open inquiry, because we believe that, especially in this type of research, it is a delicate path to gain proficiency. This is however a complex process, since the three learning domains, cognitive, inter-personal, and intra-personal are influencing each other. Findings in literature show that tasks which are too complex can frustrate students (Kirschner et al., 2006), that they can become resistant (Prince & Felder, 2007) or that the task can negatively influence their motivation (Bransford, Brown, & Cocking, 2000). In our research we found a relationship between complexity of tasks and the amount of learning in other domains. The observation that learning in one domain stimulates learning in another domain was also found in the literature. Fink discusses a taxonomy for significant learning experiences in which he describes six domains – foundational knowledge, application, integration, human dimensions, caring and learning how to learn. He emphasizes that achievement in one domain of learning can stimulate learning in other domains (Fink, 2003). Hardy, Day, Hughes, Wang, and Schuelke (2014) suggest that "in future research there is a need for greater empirical focus on the dynamics of cognitive, motivational, emotional, and behavioral self-regulatory process in active training." We think that these insights would also be supportive for training in interdisciplinary research.

We argue that when students are challenged at the right level with the appropriate guidance, the learning domains will reinforce each other. However, the role of teacher is crucial in keeping this balance. Various authors address the challenges for teachers to conduct inquiry based teaching (Quigley, Marshall, & Deaton, 2011). Zion, Schanin, and Shmueli (2013, p. 696) recognize that "the shift from teachers who direct students in a structured inquiry to those who facilitate students in an open inquiry is a challenging professional endeavor and requires greater flexibility on the part of both teachers and students." We argue that this is an even greater endeavor if the teacher has to direct students from structured inquiry to open inquiry as part of a single process.

NOTE

1. Anderson and Krathwohl (2001).

REFERENCES

Anderson, L. W., & Krathwohl. (Eds.). (2001). *A Taxonomy for learning, teaching, and assessing: A revision of bloom's taxonomy of educational objectives*. New York, NY: Longman.

Banchi, H., & Bell, R. (2008). The many levels of inquiry. *Science and Children*, *46*(2), 26–29.

Barron, B., & Darling-Hammond, L. (2008). Teaching for meaningful learning a review on inquiry and research based learning. In L. Darling-Hammond, B. Barron, D. Pearson, A. H. Schoenfeld, E. K. Stage, T. D. Zimmerman, G. N. Cervetti, & J. Tilson (Eds.), *Powerful learning what we know about teaching for understanding*. San Francisco, CA: Jossey-Bass.

Barrows, H., & Tamblyn, R. (1980). *Problem-based learning: An approach to medical education*. New York, NY: Springer.

Berg, B. L. (2004). *Qualitative research methods for the social sciences* (pp. 204–208). Long Beach, CA: California State University.

Bransford, J. D., Brown, A. L., & Cocking, R. R. (Eds.). (2000). *How people learn: Brain, mind, experience, and school.* Washington, DC: National Academy Press.

Bruner, J. S. (1961). The art of discovery. *Harvard Educational Review*, *31*(1), 21–32.

Bunders, J. F. G., Broerse, J. E. W., Keil, F., Pohl, C., Scholtz, R. W., & Zweekhorst, M. B. M. (2010). How can transdisciplinary research contribute to knowledge democracy? In R. In 't Veld (Ed.), *Knowledge democracy Consequences for science politics and media* (pp. 125–152). Dordrecht: Springer.

Dewey, J. (1933). *How we think* (p. 228). Boston, MA: D.C. Health & Co. Retrieved from http://www.archive.org/details/howwethink000838mbp

Elrod, P. D., & Tippett, D. D. (2002). The "death valley" of change. *Journal of Organizational Change Management*, *15*(3), 273–291. doi:10.1108/09534810210429309

Farrell, J., Moog, R. S., & Spencer, J. N. (1999). A guided inquiry chemistry course. *Journal of Chemical Education*, *76*, 570–574.

Fink, L. D. (2003). *Creating significant learning experiences: An integrated approach to designing college*. San Francisco, CA: Jossey-Bass.

Hardy III, J. H. H., Day, E. A., Hughes, M. G., Wang, X., & Schuelke, M. J. (2014). Organizational behavior and human decision processes exploratory behavior in active learning: A between- and within-person examination. *Organizational Behavior and Human Decision Processes*, *125*(2), 98–112. doi:10.1016/j.obhdp.2014.06.005

Kirschner, P. A., Sweller, J., & Clark, R. E. (2006). Work: An analysis of the failure of constructivist, discovery, problem-based, experiential, and inquiry-based teaching. *Educational Psychologist*, *41*(2), 75–86.

Klein, J. T., Grossenbacher-Mansuy, W., Häberli, R., Bill, A., Scholz, R. W., & Welti, M. (Eds.). (2001). *Transdisciplinarity: Joint problem solving among science, technology, and society. An effective way for managing complexity*. Basel: Birkauser.

Kolb, D. A., & Fry, R. (1975). Toward an applied theory of experiential learning. In C. Cooper (Ed.), *Theories of group process* (pp. 33–57). London: Wiley.

NVAO. (2012). Rapport beperkte opleidingsbeoordeling WO master management policy analysis and entrepreneurship in the health and life sciences vrije universiteit Amsterdam, pp. 1–28.

Piaget, J. (1964). Part 1: Cognitive development in children: Piaget development and learning. *Journal of Research in Science Teaching*, *2*, 170–186. doi:10.1002/tea.3660020306

Prince, M., & Felder, R. (2006). Inductive teaching and learning methods: Definitions, comparisons, and research bases. *Journal of Engineering Education*, *95*(2), 123–138.

Prince, M., & Felder, R. (2007). The many faces of inductive teaching and learning. *Journal of College Science Teaching*, *36*(5), 14–20. doi:10.1.1.63.7316

Quigley, C., Marshall, J. C., & Deaton, C. C. M. (2011). Challenges to inquiry teaching and suggestions for how to meet them. *Science Educator*, *20*(1), 55–61.

Savery, J. R. (2006). Overview of problem-based learning: Definitions and distinctions. *Interdisciplinary Journal of Problem-Based Learning*, *1*(1), 9–20. doi:10.7771/1541-5015.1002

Spronken-Smith, R. (2007). Inquiry-based learning: Meaning, theoretical basis and use in higher education, pp. 1–17.

Tynjälä, P. (1999). Towards expert knowledge? A comparison between a constructivist and a traditional learning environment in the university. *International Journal of Educational Research*, *31*, 357–442.

Vygotsky, L. (1978). Readings on the development of children. In *Mind and society* (pp. 79–91). Cambridge, MA: Harvard, University Press.

Wang, C., & Burris, M. A. (1997). Photovoice: Concept, methodology, and use for participatory needs assessment. *Health Education and Behavior*, *24*(3), 369–387. doi:10.1177/1090019819702400309

Woods, D. R. (2014). Problem-oriented learning, problem-based learning, problem-based synthesis, process oriented guided inquiry learning, peer-led team learning, model-eliciting activities, and project-based learning: What is best for you? *Industrial & Engineering Chemistry Research*, *53*(13), 5337–5354. doi:10.1021/ie401202k

Zion, M., Schanin, I., & Shmueli, E. R. (2013). Teachers' performances during a practical dynamic open inquiry process. *Teachers and Teaching*, *19*(6), 695–716. doi:10.1080/13540602.2013.827457

APPENDIX: TOPIC LIST FOR INTERVIEWS WITH STUDENTS AND LECTURERS OF THE MASTER PROGRAM MANAGEMENT POLICY ANALYSIS AND ENTREPRENEURSHIP

Topic list for students:

- Could you describe the learning objectives of the course?
- What new insights did you obtain during the course?
- Which elements of the course did you experience as most beneficial?
- What were the difficulties encountered during the course?
- What is different from the way of teaching if you compare it to the teaching during your bachelor program?
- Show and explain the taxonomy of Bloom.[1] What is the level of understanding of the conceptual frameworks?
- To what extent could you use the obtained knowledge and skills from the previous course(s) during this course?

Topic list for *lecturers*:

- *Students need to go through various levels of inquiry (from confirmation inquiry towards open inquiry) to be prepared to be an interdisciplinary researcher.*
  - *Why is it important that students go through the different levels of inquiry?*
- *Which level of inquiry is applied in this course?*
- *What did you experience as most challenging in the design and structuring of the course?*
- *What did you experience as challenging in teaching of the course?*
- *What are, in your perspective, the most beneficial aspects of this course?*
- *In your perspective: what difficulties did students experience during the course? (difficulties at the cognitive, intra- and inter-personal level)*
- *To what extent are students able to work on the level of open inquiry during their internship?*
  - *What are, in your perspective, the most beneficial aspects of the internship?*
  - *In your perspective: what difficulties did students experience during their internship?*

Table A1. Learning at the Three Domains of Learning at the Various Levels of Inquiry (described by Banchi & Bell, 2008).

	Confirmation	Structured	Guided	Open
Cognitive	Practice application of interdisciplinary research in a sandbox environment.	Collect, interpret, and integrate data in order to answer the research question.	Specify the research question and collect, and integrate data to develop a plan.	Formulate the research question, select framework and methodology, collect, analyze, and compare data in order to answer the research question.
Inter-personal	Acceptance, deal with uncertainty, respect, value, appreciation.	Listen actively, communicate effectively, demonstrate leadership.	Conflict management, negotiation, convince, appreciate different points of view.	Work in a dynamic, social environment.
Intra-personal	Motivation, confirmation of self.	Adaptability, acceptance, (self) reflection.	Tolerance for stress, ambiguity, critical assessment of own role in team.	Be afraid but do it anyway.

A NARRATIVE-BASED BACHELOR OF SCIENCE IN NURSING PROGRAM: EXPERIENCES AND INSIGHTS

Baiba Zarins and Lorraine Carter

ABSTRACT

Nipissing University in North Bay, Ontario, Canada, working in partnership with leading health science centers in Toronto, Ontario, has developed a unique second-degree entry Bachelor of Science in Nursing (BScN) program called the Scholar Practitioner Program. This program is a response to the increasing demand for nursing programs which builds on prior university learning.

The program uses an educational model based on narrative inquiry and cognitive apprentice pedagogies. Narrative inquiry pedagogy is the overarching philosophical framework of the program, which embraces values that connect teachers and learners. A spirit of inquiry is cultivated in every student and a research culture embedded in the student's practice.

The six-semester two-year full-time program occurs in a learner-driven environment which shapes how specific semester program objectives are

Inquiry-Based Learning for Science, Technology, Engineering, and Math (STEM) Programs: A Conceptual and Practical Resource for Educators
Innovations in Higher Education Teaching and Learning, Volume 4, 193–216

ISSN: 2055-3641/doi:10.1108/S2055-364120150000004012

met. A strong emphasis is placed on experiential learning within the Toronto-based academic health sciences centers.

A continuous interactive process involving teachers and learners encourages self-directed learning and participant accountability. Application of knowledge and skills in a professional, caring, and holistic manner is expected. This type of undergraduate learning environment which includes immersion in the employment milieu enables the future scholar practitioner to be relevant in the evolving profession of nursing.

INTRODUCTION

Health care careers, particularly in nursing, are attractive because of the promise of ongoing employment. Our societal demographics provide evidence of the need for an enhanced and responsive workforce to meet the growing and increasingly complex health care needs of North American and global populations. According to workforce analysts, nursing shortfalls in the United States will surpass 500,000 by the year 2025 (Buerhaus, Staiger, & Auerbach, 2008). Although the supply of regulated nurses in Canada continues to outpace population growth, nursing demographics are shifting within Canadian markets (Canadian Institute for Health Information, 2014). Nurses of the boomer generation are retiring while the demand for nurses in preventative health and health system navigation along the continuum of care – from cradle to grave and including wellness and illness – is increasing. As a result, "[n]ursing is ranked as one of the most promising careers in the coming decades and most schools are oversubscribed. We need to think strategically and creatively when it comes to overcoming the challenges in our health system and recognize that educational institutions are part of the solution to meeting the demand for health services" (Canadian Association of Schools of Nursing, 2008). According to the Canadian Nurses Association (2013), fast-track programs are becoming increasingly popular and more accessible to students as evidenced by the almost 19% growth in such programs in 2006-2007 to 48.6% in 2013. They are seen as part of the solution of increasing the nursing supply as they enable students to complete their education in less time than traditional programs and include accelerated, compressed, second-degree entry, advanced entry, and bridging programs.

While every province in Canada now offers multiple fast-track programs in nursing, the admission requirements and the flexibility of these programs vary. As an example, second-degree nursing programs offered by Canadian universities vary in admission requirements and program structure. The Bachelor of Science in Nursing (BScN) program described in this chapter builds on prior university learning and uses a pedagogical framework emphasizing narrative inquiry and learning in the point of care setting. This immersion of the student in point of care contexts is crucial to learning the nursing role in health care delivery. Juxtaposed with other second degree nursing programs in Ontario, the Scholar Practitioner Program is the first of its kind designed over six semesters and serving students from diverse learning backgrounds. Students enter the program with the required undergraduate degree in a variety of disciplines, including the natural sciences, the social sciences, and the arts as well as internationally conferred degrees. The program is learner focused in its design, curriculum, learning modalities, and overall context.

Today, nursing education is required to prepare a workforce that can practice effectively in rapidly changing, technologically defined health care environments (National League for Nursing, 2003). Nurses are expected to be self-directed and ready to apply evidence-based knowledge in practice (Canadian Association of Schools of Nursing, 2005). In order to achieve this, nursing programs may need to move beyond traditional curricula and embrace a significant paradigm shift focusing on innovation (National League for Nursing, 2003). While students need to bring their prior knowledge and experiences to their nursing studies, new pedagogies that are research-based and responsive to a rapidly changing health care system require self-directedness, academic maturity, and comfort with a high level of independence in study. Additionally, these pedagogies require new collaborative partnerships within the practice environment, new ways of learning, and contemporary technologies (National League of Nursing). By doing this in the Scholar Practitioner Program, the course of study fosters nurse graduates that will be "practice-ready scholar practitioners" possessing the knowledge, skill, judgment, and ethical components required to anticipate the care needs of patients/clients/families/communities along the health continuum. Graduates will also have the ability to mobilize resources on their behalf.

PROGRAM DESIGN

Similar to other second degree program models, students can complete their Bachelor of Science in Nursing (BScN) in two calendar years. What is different in this program is that faculty mentor students as co-learners within a narrative inquiry-cognitive apprenticeship-based curriculum. Narrative inquiry is the overarching philosophical framework of the program and affords teachers and learners access to the human experience of time, order, and change. It also obligates us to listen to the human impulse to tell stories (Carr, 1986; Connelly & Clandinin, 1990; Crites, 1971; Mishler, 1990; Polkinghorne, 1988). In cognitive apprenticeship, how knowledge translation intersects with the preceptor-learner-patient-client-faculty relationship is important. The methods of cognitive apprenticeship include modeling, coaching, scaffolding, articulation, reflection, and exploration (Dennen & Burner, 2002; Stalmeijer, Dolmans, Wolfhagan, & Scherpbier, 2009). Narrative inquiry pedagogy within a cognitive apprenticeship model situates learning as stories told by clients, families, teachers, practitioners, and clinical preceptors. These stories are experienced and interpreted by students within learning groups. Faculty act as consultants, not lecturers; as learners, they generate and share stories and cases within various paradigms (Swenson & Sims, 2000).

The program has a strong emphasis on experiential placements within academic and research-oriented health care delivery organizations. The outcome for the student is immersion within the context of health care provision at the point of care. Each affiliate organization provides learners with substantive breadth and depth in point of care learning experiences along the continuum of care trajectory. Learners observe and work within the context of care, shaping meanings, understandings of the overall health care system, and the multidimensional role of nursing within this landscape. This milieu comprises a clinical experience opportunity for inquiry-led evidence-based active learning. Students as co-learners seek out the narratives of patients, clients, and families in generating further questions for consideration and learning.

As a sequential six semester, full-time program, the program unfolds in a learner-driven environment which is based on the students' foundational academic studies. Many students have experienced more traditional pedagogies and welcome the uniqueness of the program's approaches to learning. Faculty from the affiliate health organizations mentor students to create an environment of discovery within the program. Scheduled reflection sessions provide co-learners opportunity to engage in sharing, researching, and discovering new ways of knowing. Important to this process is changing how we define educational moments. The activities and measures of traditional pedagogies do not always fit the practice environment nor the opportunities and realities experienced by individual learners.

This program, however, offers independent, mature, and motivated adult learners a venue to attain undergraduate entry to practice nursing education grounded in a learner-centered curriculum.

PROGRAM CURRICULUM

This program is grounded in an educational model based on narrative inquiry and cognitive apprentice pedagogies which must meet accreditation standards as well as provincial and national entry to practice regulatory requirements for licensure with the College of Nurses of Ontario (College of Nurses of Ontario, 2014). The act of inquiry into the core experience of the client serves as a basis for initial communication. Such communication can be significantly valuable to initially acquire the patient's medical history, but more importantly in understanding the patient/client/community story within their health care context and larger dialogue towards goal setting through mutual collaboration.

Narrative inquiry pedagogy focuses on the processes of teaching and interpreting as well as critical thinking and analysis of concepts, ideas, and situations. This method delves deeply into persons' narratives and, when applied to knowledge acquisition, fosters the creation of learning partnerships involving students, faculty, and preceptors rather than traditional teaching and learning hierarchies. As suggested by Lindsay (2006) and Lindsay and Smith (2003), the primary resource of the faculty educator is his or her experience of being a nurse and the understanding that nursing education is a mutually enlightening role. This approach adds to the view held by Clandinin (2006) that faculty knowledge is exclusive "storied life compositions" that reflect life history and social milieu to be transitioned to student. According to this understanding, story holds fundamental importance within any context of learning for all participatory learners.

Curriculum development in this program model has involved knowledge from a variety of disciplines. The natural sciences, social sciences, and human sciences contribute to the development of nursing knowledge and constitute a broad basis for learners to understand the context of the human health experience. Faculty and learners explore content, nursing concepts, and nursing issues from a variety of perspectives. Both quantitatively and qualitatively constructed knowledge is valued. From a quantitative perspective, learners are required to achieve pre-established milestones and levels through structured adaptive learning examinations and simulated point of care case scenario testing. Qualitatively, students express translation of nursing knowledge and achievement of their learning plan goals through narrative reflections, essay assignments, learning portfolios, and interactive presentations. Multimodal learning approaches include aesthetic artefact activities; one such example will be described later in this chapter. E-strategies along with infrastructure resources, such as interprofessional education workshops, seminars, and point of care dialogues offered by the affiliate organizations, are also valued. Immersive learning strategies are most often utilized by learners in meeting their individual goal pathways in achieving semester-based objectives. In collaboration with program faculty and preceptors, learners set specific goal objectives each semester for subsequent formative and summative assessment by all stakeholders.

INQUIRY-SITUATED LEARNING IN ACTION

In his studies of adult learning, Knowles (1984) observed that adult students approach learning within the context of life application. In the case of the Scholar Practitioner Program, this pedagogical philosophy is part of the attraction of the program. In brief, the program meets individuals' diverse learning styles within situational contexts. Awareness of self and reality is critical. Placing the learner at the intersection of the learning environment (context) and knowledge application within the point of care environment fosters a spirit of inquiry and discovery into the health and life experience of the learner, as well as the professional relationship of patient/client/family/community through which interactive learning occurs.

Inquiry and discovery are equally important among the faculty. Below are unedited notes taken during a faculty development session. In their raw form of questions, statements, and partially articulated ideas, these notes reveal the different ways of thinking, learning, and problem solving of the faculty. This diversity is vital to fostering inquiry and discovery among the students as learners:

> *How do I learn?* Exploration of learning styles and cognitive psychology. Utilization of inquiry and reflection as learning tools. Composing a narrative of historical learning and factors contributing to one's individual and intersectional learning journey. The overall goal is a realistic understanding and creation of a learning assessment. A related activity includes integration of learning styles and narratives into individualized learning plans and portfolios.

How is knowledge created? Discussion of divergent viewpoints of knowledge. How does one define knowledge? Reflection of self in learning context (social critical theory, privilege of education).

How is competence defined? What are the social, economic and cultural distinctions that define competence? How does professional competence juxtapose with the nursing metaparadigm which informs the foundational and philosophical practice of nursing?

How will I achieve program outcomes? Review of program objectives and two year, six semester themes in relation to graduate outcomes. Learners explore the various methodologies in achieving their learning goals. How will I get there? Mapping of competencies by learner throughout individual learning plans, semester objectives and professional competency standards.

How am I a professional? (The inquiry of self) How do I define the criterion of my profession? This subsequent question explores concepts surrounding professional ethics and accountability.

How am I in a narrative way? (Reflection of self in action) How does ones narrative lead towards a profession in nursing? The leading question is posed to share lived experiences in accountability and professionalism.

How are the two concepts above linked? (Experiential intersect) What are the practicum experiences that shape my professional identity and self? (The leading question to further demonstrate knowledge acquisition, knowledge translation and knowledge networking with the learning context).

Using inquiry-based questions of discovery, learners and faculty participate in evidence-informed discourse to integrate resulting new knowledge and its adaptation into professional practice learning. This methodology is introduced, implemented in practice, and re-introduced with increasing breadth and complexity throughout the two years of study. This approach to increased complexity or leveling over time strengthens foundational understandings and builds on existing knowledge to achieve enhanced and focused integration of the narrative within the learning environment.

Within the first year of study, learners discover and explore different concepts, content, and types of learning (cognitive, affective, and psycho-motor skills/multi-skills) in order to acquire basic knowledge, introductory skills, and beginner confidence in relation to course and program outcomes. Concepts are introduced, discussed, and re-introduced in settings of increasing complexity and facilitate repeated exposure to the nursing competencies that comprise the entry to practice standards of nursing's regulatory body.

Explicit detailing of curriculum expectations is outlined within the student handbook. Program expectations are used to shape individualized learning plans and formative skill development through multimodal approaches to learning. An excerpt from the Student Handbook (Nipissing University, 2014) below illustrates the program's overarching curricular components:

- Theoretical foundations
- Standards and entry to practice competencies
- Exploring and understanding human health assessment
- Understanding communication interviews and therapeutic communication
- Exploring diversity and cultural care competency
- Language and concepts for care and practice
- Nursing process
- Concepts and theories in nursing
- Concepts theories and models in health care and systems care delivery
- Patient-client- and family-centered care (partners in care).

Each of the above elements, as presented in the SPP handbook, is identified within the context of increasing complexity across the six semesters to foster progression in the inquiry of learning, creativity among learning modalities, and achievement of terminal outcomes and learning milestones. The curriculum is continuously reviewed, enhanced, and adapted by the faculty to fit the needs of the learners as well as to reflect the contextual learning environment in which core competency requirements are met. Given this process, learning is, as Clandinin, Pushor, and Orr (2007) point out, an inward, outward, forward, and backward relationality within the multidimensional and changing contextual learning environments of health care. This approach further reflects the complex dimensions of narrative inquiry pedagogy as it includes inquisitiveness, attentiveness, self-exploration, and mindful interactions. Table 1 presents the learning domains and concepts found in the program. It further points out key distinctions related to leveling between years 1 and 2, pointing out that, at all stages in the program, the student is expected to participate in activities of introduction and discovery (I), exploration and emphasized inquiry (E), reinforcement and critical reflection (R), and applied and reflective practice (A). Table 2 presents the general rubric and criteria used to determine the level or degree of complexity of specific outcomes.

Table 1. Curriculum Development Concepts and Co-Learner (Student) Learning Leveling Milestones.

Learning Domains & Concepts	*SELF-DIRECTED & REFLECTIVE PRACTICE LEARNING OPPORTUNITIES FOR BUILDING KNOWLEDGE & ENTRY TO PRACTICE COMPETENCIES (All areas of learning & reflective practice will reflect peer reviewed and/or scholarly research – EBP sources when possible)*
	The Level of Instruction Utilizing Narrative Inquiry
	The leveling of instruction refers to the level and complexity of the program outcome – related knowledge and skills learning expected to be "*taught within narrative inquiry pedagogy*" and learned in Year 1 and Year 2 Semester courses of the SPP program.
	Introduction & Discovery (I)
	1. Co-learners (students) are not expected to be familiar with the program outcome-related concept/content or skills at the baccalaureate level. 2. Instruction, guided discovery and learning activities focus on basic knowledge, skills, and/or competencies and the beginner entry-level complexity. 3. Only one or a few aspects of a complex program outcome are addressed in the given course.
	Explored & Emphasized Inquiry (E)
	1. Co-learners (students) are expected to possess a basic level of program outcome-related knowledge and familiarity with the concept/content or skills at the baccalaureate level. 2. Instruction, guided exploration and learning activities concentrate on enhancing and strengthening knowledge, skills, and expanding complexity. 3. Several aspects of the program outcome are addressed in the given course, but these aspects are treated separately.
	Reinforced & Critically Reflected (R)
	1. Co-learners (students) are expected to possess an advanced level of the program outcome related knowledge, skill, or competency at the baccalaureate level. 2. Instructional and learning activities focus on the use of the concept/content or skills in multiple contexts and at multiple levels of complexity. 3. Given program outcome is addressed in all of its complexity across multiple contexts or is turned reflexively on itself.

Table 1. (*Continued*)

Applied & Reflective Practice (A)

1. Co-learners (students) are expected to possess an advanced level of program outcome-related knowledge, skill, confidence and mastery (competency) at the baccalaureate/entry to practice competency level.
2. Instructional and learning activities focus on the use of the content or skills in multiple contexts and at multiple levels of complexity.
3. Given program outcome is addressed in all of its complexity across multiple contexts or is turned reflexively on oneself.

As adapted & modified from: Cuevas & Feit (2011). *Curriculum Mapping: A Roadmap for Curriculum Coherence and Student Achievement*, Southern Connecticut State University, March 4, 2011. Nuria M. Cuevas (ncuevas@nsu.edu), Marvin D. Feit (mdfeit@nsu.edu)

Year 1 – Leveling & Type of Learning

The level and type of learning (I), (E), and initial use of (R) and (A) are the expected self-directed learning outcomes of the Year 1 co-learner. In this description, (I), (E), (R), and (A) refers to the level (modified from Rubric to determine levels of program outcome content delivery in courses as found below) and the complexity of the program outcome-related knowledge and skills that are expected to be developed through content/concepts discovered (introduced), explored (emphasized), and reflected on/reinforced as a new beginner/learner in Year 1 courses of the program. Therefore, in Semester 1, 2, and 3, the co-learner is progressing from the introduction and discovery of new concepts and content knowledge to that of further explored and emphasized inquiry at the beginner level of new knowledge. The beginner level student in Year 1 may also exceed this to continue to promote learning of specific concepts and content beyond Year 1 expectations including that of being able to reflect a more fulsome awareness, breadth of curiosity and depth of complexity of content/concept topics being considered, explored, deconstructed and reconstructed as a way of knowing and of determining meaning for the context of self. As a result of this approach in learning, co-learners may embrace and exhibit use of all four levels of concept/content delivery within Year 1– illustrating the full narrative inquiry paradigm experience which would include: Introduced/Discovery (I), Explored/Emphasized Inquiry (E), Reinforced/Critically Reflected (R), and Applied/Reflective Practice (A). The levels are enhanced or further modified from those developed and defined by Cuevas & Feit (2011). *Curriculum Mapping: A Roadmap for Curriculum Coherence and Student Achievement*, Southern Connecticut State University, March 4, 2011. Nuria M. Cuevas (ncuevas@nsu.edu), Marvin D. Feit (mdfeit@nsu.edu).

Thus, in Semesters 1, 2, and 3 of the SPP courses and Year 1 of the program, co-learners begin discovering and exploring (as shared inquiry) concept/content level and type of learning (cognitive, affective and psychomotor skill/multi skills) at the basic, new learner, beginner level of knowledge, skills, and confidence and mastery (competency) for course/program outcomes.

Year 2 – Leveling & Type of Learning

The level and type of learning (I), (E), and enhanced proficiency of use of (R) and (A) are the expected self-directed learning outcomes of the Year 2 co-learner. In this description, (I), (E), (R), and (A) refers to the level (modified from Rubric to determine levels of program outcome content delivery in courses as found below) and the complexity of the program outcome-related knowledge and skills that are expected to be developed through content/concepts discovered (introduced), explored (emphasized), and reflected on/reinforced and applied/reflective practice as an advanced beginner learner in Year 2 courses of the program. Therefore, in Semesters 4, 5, and 6 the co-learner is progressing from the introduction and discovery of new concepts and content knowledge to that of further explored and emphasized inquiry at the advanced beginner level of now new and enhanced knowledge. The advanced beginner level student in Year 2 may also exceed this to continue to promote learning of specific concepts and content beyond Year 2 expectations including that of being able to reflect a more fulsome awareness, breadth of curiosity and depth of complexity of content/concept topics being considered, explored, deconstructed and reconstructed as a way of knowing and of determining meaning for the context of self. As a result of this approach in learning, co-learners may embrace and exhibit use of all four levels of concept/content delivery within Year 2– illustrating the full narrative inquiry paradigm experience which would include: Introduced/Discovery (I), Explored/Emphasized Inquiry (E), Reinforced/Critically Reflected (R), and Applied/Reflective Practice (A) and exhibiting strength in complex problem solving; critical thinking and reasoning; a well-developed sense of curiosity; and critical scholarly inquiry as a support to evidence-based decision-making.

Thus, in Semesters 4, 5, and 6 of the SPP courses and Year 2 of the program, co-learners continue developing knowledge (depth and breadth) and demonstrate enhanced comprehension of concept/content level (complexity) and enhanced proficiency in type of learning (cognitive, affective and psychomotor skill/multi skills) which supports more advanced knowledge, depth of concept understanding, and the emergence of the more advanced beginner confidence and mastery (competency). This level of development is the knowledge development and progression in competency leveling expected in both the courses and program outcome required at completion of Semester 6. As a result, in Semester 6, co-learners are able to demonstrate advanced beginner knowledge, skill and judgment (critical thinking and reasoning) and meet all College of Nurses, Entry to Practice level competencies.

Year 1 – Discover, Explore, & Emphasize Curious Inquiry & Reflect.

Year 2 – Discover, Explore, & Emphasize Curious Inquiry, Critically Reflect & Applied (Reflective Practice).

(Also see Year 1 & Year 2 detailed Mapping – Curriculum Development Concepts and Co-Learner Leveling Milestones Documents).

Table 2. A Rubric to Determine Levels of Program Outcome Content Delivery in Courses (I, E, R, A).

Levels of Program Outcomes Content Delivery	General Factors – Defining Course Level of Content Delivery in the Context of the Program Outcome Content Domain	Course Focus – in the Context of the Program Outcome Content Domain (Plaza et al.)	Focal Cognitive Behaviors – In the Context of the Program Outcome Content Domain (Bloom, Anderson et al.)	Action Verbs – in the Statements of Course Learning Outcomes/ Assessment Tasks Related to the Program Outcomes (Biggs)	Student Intellectual Tasks – in the Context of the Program Outcome Content Domain (Knefelkamp)
Introduction & Discovery (I)	1. Students are not expected to be familiar with the program outcome related content or skill at the baccalaureate level. 2. Instruction and learning activities focus on basic knowledge, skills, and/ or competencies and entry-level complexity. 3. Only one or a few aspects of a complex program outcome are addressed in the given course.	An indirect relationship exists between the course and the program outcome. In this case, the given program outcome itself is not the focus of the course, but at least one element of the course serves as a building block to the achievement of the given program outcome	*Remembering*: Retrieve relevant knowledge from long-term memory by: • Recognizing • Recalling	Understanding of the material related to the given program outcome is nominal • Identify • Recognize • Define • Paraphrase • Choose • Select • Calculate • Arrange • Find • Follow (simple) instructions	Learning basic information and definitions of terms and concepts. Learning to identify parts of the whole within the context of the program outcome. Beginning to be able to compare and contrast things.
Explored & Emphasized Inquiry (E)	1. Students are expected to possess a basic level of program outcome related knowledge and familiarity with the content or skills at the baccalaureate level.	A more direct relationship exists between the course and the program outcome.	*Understanding*: Construct meaning from instructional messages, including oral, written, and graphic communication by:	Understanding of the material related to the given program outcome as *"knowing about"* • Describe • Account for • Classify	Can do compare-and contrast tasks. Can see multiples – perspectives, parts, opinions, and evaluations. Perform basic analytic tasks.

	2. Instruction and learning activities concentrate on enhancing and strengthening knowledge, skills, and expanding complexity. 3. Several aspects of the program outcome are addressed in the given course, but these aspects are treated separately.	A mixture of course elements supports the achievement of the given program outcome, but the final integration of the knowledge, skills, and attitudes necessary for its achievement may not be accomplished in this course.	• Interpreting • Exemplifying • Classifying • Comparing • Inferring *Applying*: Carry out or use a procedure in a given situation by: • Executing • Implementing	• Structure • Formulate • Execute • Solve • Prove • Do algorithm • Apply method	Use supportive evidence.
Reinforced & Critically Reflected (R)	1. Students are expected to possess an advanced level of the program outcome-related knowledge, skill, or competency at the collegiate level.	A direct relationship exists between the course and the program outcome.	*Analyzing*: Break material into its constituent parts and determine how the parts relate to one another and to an overall structure or purpose by: • Differentiating • Organizing • Attributing	Understanding of the material related to the given program outcome as *"appreciating relationships"* • Analyze • Explain • Compare • Contrast • Integrate • Summarize • Design • Relate • Explain causes • Apply theory (to its domain)	Good at analysis. Able to critique with positives and negatives.
	2. Instructional and learning activities focus on the use of the content or skills in multiple contexts and at multiple levels of complexity. 3. Given program outcome is addressed in all of its complexity across multiple contexts or is turned reflexively on oneself.	At least one element of the course focuses specifically on the complex integration of knowledge, skills and attitudes necessary to perform the given program outcome.			Use supportive evidence well. Can relate learning to other issues in other classes or to issues in "real life" – if they will apply themselves to that task. Learning to think in abstractions.
Applied & Reflective Practice (Advanced) (A)	1. Students are expected to possess an advanced level of program outcome-related knowledge, skill, or	A direct relationship exists between the course and the program outcome.	*Evaluating*: Make judgments based on criteria and standards by: • Checking • Critiquing	Understanding of the material related to the given program outcome as *'far transfer'*, that is the ability to generalize to	Can evaluate, conclude, and support own analysis. Can synthesize. Can adapt, modify and expand concepts

Table 2. (*Continued*)

Levels of Program Outcomes Content Delivery	General Factors – Defining Course Level of Content Delivery in the Context of the Program Outcome Content Domain	Course Focus – in the Context of the Program Outcome Content Domain (Plaza et al.)	Focal Cognitive Behaviors – In the Context of the Program Outcome Content Domain (Bloom, Anderson et al.)	Action Verbs – in the Statements of Course Learning Outcomes/ Assessment Tasks Related to the Program Outcomes (Biggs)	Student Intellectual Tasks – in the Context of the Program Outcome Content Domain (Knefelkamp)
	competency at the collegiate level. 2. Instructional and learning activities focus on the use of the content or skills in multiple contexts and at multiple levels of complexity. 3. Given program outcome is addressed in all of its complexity across multiple contexts or is turned reflexively on oneself.	The course primarily focuses on the complex integration of knowledge, skills and attitudes necessary to perform the given program outcome.		novel situations, and as involving meta cognition: • Discuss • Assess • Evaluate • Theorize • Generalize • Hypothesize • Predict • Judge • Reflect • Transfer theory (to new domain) There is use of the content or skills in multiple contexts and at multiple levels of complexity.	because they understand the concepts. Relate learning in one context to learning in another with some ease. Look for relationships in the learning and addresses in all of its complexity across multiple contexts or is turned reflexively on oneself. Can critically think and reason in the context of complexity

Source: Modified from Cuevas & Feit (2011).
Nuria M. Cuevas (ncuevas@nsu.edu), Marvin D. Feit (mdfeit@nsu.edu).

WHAT ARE THE IMPLICATIONS OF NARRATIVE-BASED LEARNING FOR EDUCATORS AND LEARNERS?

The teaching and learning philosophy for the program embraces values and beliefs that connect teachers and learners and informs how learning occurs. Core philosophical elements influence the design of the program and expectations in teaching and learning experiences. Every effort is taken by faculty and preceptors to cultivate a spirit of inquiry among learners and to embed a research culture in practice. Learning plans and portfolios including artefacts are prepared each semester. Both plans and portfolios are based on evidence-informed practice and support learners in achieving required educational outcomes. This approach to learning is what is required to produce "scholar practitioners" for contemporary nursing practice. Further, these philosophical underpinnings emerge from a human sciences perspective and, as such, lead to understanding the human experience of health in deeply personal, historical, experiential, and contextual ways. Significantly, the human sciences including language and visual arts offer unique experiences of self-reflection that reveal the human condition in ways not possible through more structured disciplines.

Ideally, education is an interactive process between teachers and learners within an environment that encourages self-directed learning and accountability of all stakeholders. These characteristics and features are central to the Scholar Practitioner Program. Uniqueness, open inquiry, and a commitment to continuing education are promoted.

Through the partnership with the academic health centers, a gateway to research, knowledge, and nursing education is established for the learners who create valuable connections and experience research in their clinical learning. This partnership provides learners with rich learning experiences and prospective employment opportunities within the participating organizations. Preliminary graduate data reveals that over 90% of graduates are offered employment within their health care delivery organizations and in programs of interest. An area of future research includes longitudinal retention rates in relation to existing provincial new graduate employment studies.

In a narrative inquiry curriculum, co-learners including faculty and students place themselves within a social context; they become deeply interested in exploring both relationships and scholarship. Similarly, they are co-constructors of knowledge who come to understand what they are experiencing and how their emerging knowledge is multi-dimensional in experience and end result. On many levels, this continuous interaction creates extensive overlapping professional relationships in learning.

Based on the work of Schon (1983) and Schwab (1973), the framework for the program curriculum is multidimensional and complex. It is founded on the concepts of learner, adjunct professor, milieu (as environment including preceptor and client), and subject matter (as content). The dimensions of the framework are connected: each learning moment is intended to capture multiple concepts along each interface. This way, the learning moment and experience add depth and meaning as well as relevance to the student's developing practice. Nursing is a practice profession; thus, the purpose of all learning experiences is to contribute to and enhance professional practice.

INQUIRY-BASED EXAMPLES AND APPLICATIONS

While more traditional forms of nursing education are strongly based in empirical knowledge, equally important is aesthetic knowledge as explained by Carper (1978, 1978) who identifies the ways of knowing within nursing as empirical, personal, ethical, and aesthetic. Building on Carper (1978), Johns (1995) points out that aesthetic knowing requires the nurse to explore questions such as the following: What was I trying to achieve? Why did I respond as I did? What were the consequences of that for the patient? Others? Myself? How was this person(s) feeling? How did I know this?

While these understandings of nursing knowledge have existed in the nursing literature for decades, more and more, nursing programs are expanding to include multimodal and non-traditional approaches to attaining knowledge. The Scholar Practitioner Program ensures that learners come to understand nursing in its multidimensionality and as an experience of diverse ways of knowing. As an example, the concept of relational inquiry can take on multiple meanings in of the "art of nursing": it relates to skill and knowledge as well as the level of expertise of specific skills and practical knowledge base. Extrapolating, students may learn from and through the visual and literary arts. Individual interpretation and classification of art in the context of nursing fosters a philosophical problem, namely, "... what kind of activity constitutes nursing as distinct from other health care interventions, and how to distinguish good nursing care" (Wainwright, 2000, p. 753).

Within this nursing program, application of relational inquiry involves an aesthetic approach to learning. Concepts such as vulnerability, empathy, compassion, and caring can be exemplified through aesthetically oriented learning activities: creative art, photography, sculpture, and mask-making activities are examples of such activities. While learners in the program familiarize themselves with the principles and practices of relational inquiry through text readings and media presentations of storied experiences, the most powerful means is the mask-making activity conducted in the first semester of the program. This activity involves pairs of learners with one person preparing and creating a plaster mask of his or her partner.

The intimacy of this activity requires open communication, continuous dialogue to establish trust, and vulnerability. The receiver of the mask must also be open to experiencing a mask of one's own face. Through this experience, learners engage in dialogue that explore the self and the meaning of the concepts from different perspectives including that of the mask maker and the mask recipient. Because the learners have entered the program with at least one university degree and often with rich life and professional experiences, they generally have ability to reflect deeply and honestly as well as articulate their understandings of self. Reflection is situated in decorating the mask to represent both the external "face" and the internal "face" which requires self-awareness and understanding. Both are equally important in presenting oneself within the professional dialogue of self and with others. Facilitative guiding questions of relationality and caring concepts are used as prompts by faculty to engage learners in exploring themselves and others within this creative experience. This type of activity is revisited later to elicit its effect on learners once they experience their first clinical practicum phase with patients/clients/families/communities in exploring the meaning of the mask activity in the context of health care provision (Fig. 1).

Debriefing reflective journals and interactive discussions about the mask making activity situate this experience within a story and context-centered moment of learning. Students have informally remarked that the activity evokes strong memories of past experiences including unknowns, fears, assumptions, and feelings. Taken together, these elements generate deeper understanding of the patient-client interface of care, whereby

Fig. 1. Photos of Mask Making Activity.

Fig. 1. (Continued)

patients/clients/families may experience similar feelings within the health care interface. Stories are shared and interpreted by faculty and learners as experiences that assist in understanding the therapeutic nurse-client relationship.

This activity is only one that derives from the arts. Stories, film, poetry, art, and drama are all part of the lived curriculum. Students as learners seek new knowledge in a community of discovery and a culture of inquiry.

BUILDING THE INQUIRY-BASED PLATFORM THROUGH DIFFERENTIATED LEARNING MODALITIES

Inquiry-based learning involves diverse learning strategies that support discussion and interdisciplinary approaches to fostering active and collaborative discussion. A focus on situated learning inclusive of client strengths and challenges, as well as the broader social determinants of health, offers a broad spectrum of reference and considerations in formulating health care strategies or interventions. Overlaying the foundational nursing process – assessment, planning, intervention, and evaluation – further develops higher levels of therapeutic considerations shaped by the learner and the learning environment. In work by Magnussen, Ishida, and Itano (2000), the impact of inquiry-based learning on critical thinking in nursing programs was examined. The benefits for students included enhanced exposure to the critical thinking skills required in nursing practice and increased confidence levels.

INQUIRY AS DISCOVERY FOR STUDENTS AND FACULTY

Educational theorists identify three broad areas of learning: the cognitive, affective, and psychomotor (Stewart, Tondora, & Hodge, 2004). Comparatively, learning styles among health care professionals have been described as activist, reflective, theoretical, and pragmatic in approach (Grol, Bosch, Hulscher, Eccles, & Wensing, 2007; Lewis, Bolden, 1989). In the case of the lived curricular experience of the program, learners are afforded the experience of integrating all aspects of their learning and encouraged to use varied learning styles. Program participants frequently refer to this kind of freedom as "structured ambiguity." Additionally, all participants bring inquiring minds and a desire to develop the knowledge, skills, and attitudes required to practice effectively in today's complex health care environment. As co-learners, they bring life experiences, beliefs, and values that inform and influence their learning processes. As mentioned previously, the faculty bring scholarly knowledge about nursing education as well as practice experience to the program. They further bring a desire to share their knowledge with students in ways that facilitate knowledge acquisition, support development of critical thinking skills, and encourage generation of personal meaning for the students.

The essence of the nursing curriculum emerges from the interactions among (not merely between) faculty and students, all of whom are co-learners. These interactions take place in scheduled learning sessions, laboratories, clinical practice settings, and assessment situations. The participants in these interactions are students, professors, practicing nurse preceptors, other members of the health care team, and clients. The curriculum is what actually occurs within these relationships; it comes to life through purposeful exploration of the content of the program in addition to learners' realizations that they are achieving their plans to become nurses. Formative and summative assessment serves as evidence of the students' achievement of their plans. Within these interactions, the roles of professor and learner are fluid and dynamic. Each is engaged in teaching and learning, cultivating professional practice, and furthering understanding of the self.

The continuous engagement of teachers and learners encourages self-directed learning and participant accountability. Application of knowledge and skills in a professional, caring, and holistic manner is expected. Like other undergraduate nursing education experiences, the SPP offers a broad general education, preparation in professional nursing, and a basis for graduate study. Significantly, graduates of the program have demonstrated their commitment to lifelong learning in post-graduate health care studies including medicine, global health, public health, and specialized nursing fields. This disposition towards learning, positioned along the continuum of nursing education to nursing practice, is what enables the scholar practitioner to be relevant within an evolving and increasingly mobile profession.

FUTURE DIRECTIONS

Nurses of the future need to be scholar practitioners: they need to be practitioners who are client centered and who partner with patients/clients and families/communities along the continuum of health and illness and living and dying. Nurses likewise need to be self-directed leaders in best practice and champions of quality of care. Further, they must be skilled users of new technologies for the assessment and treatment of patients, clients, families, and communities in the delivery of health care services. In this program, given the sophistication of the health centers in which the students learn, they are exposed to and learn to use cutting edge health technologies on a daily basis. Additionally, the learners in the program use a cross section of new and emerging learning and communication technologies in their studies. Their competence and confidence with technology in general are vital to their forthcoming work as nurses in changing times. As networkers and relationship builders moving within the health care system, nurses must also possess competence and confidence in leading teams of professionals and support staff in the health care settings. Learners in the SPP are well advantaged in this domain too.

While there are many and different ways of achieving the above, the Scholar Practitioner Program offers a unique opportunity to second degree students who desire a different route to their goals. As a program grounded in discovery, inquiry, experiential learning, narrative as a form of representing health and wellness, and learning conceptualized as a shared journey, the Scholar Practitioner Program is an excellent choice for persons who embrace these same values. Finally, as health care becomes more complex with an aging population, fiscal restraints, and new health challenges in our global society, it will be increasingly important for the system to be distinguished by nurses who, as practitioners and leaders, are skilled problem solvers, expert thinkers, and evidence-seeking scholars.

REFERENCES

Buerhaus, P., Staiger, D. O., & Auerbach, D. I. (2008). *The future of the nursing workforce in the United States: Data, trends and implications*. Boston, MA: Jones and Bartlett Publishers.

Canadian Association of Schools of Nursing. (2005). Position statement on baccalaureate education and baccalaureate programs, Ottawa, CA.

Canadian Institute for Health Information. (2014). *Regulated nurses 2013*. Retrieved from https://secure.cihi.ca/free_products/Nursing-Workforce-2013_EN.pdf

Canadian Nurses Association. (2013). Position statement on baccalaureate education and baccalaureate programs, Ottawa.

Carper, B. (1978). Fundamental patterns of knowing in nursing. *Advances in Nursing Science*, *1*, 13–23.

Carr, D. (1986). *Time, narrative and history*. Bloomington, IN: Indiana University Press.

Canadian Association of Schools of Nursing. (2005). *Position statement on baccalaureate education and baccalaureate programs*. Ottawa, CA.

Clandinin, D. (2006). Narrative inquiry: A methodology for studying lived experience. *Research Studies in Music Education*, *27*, 44–54.

Clandinin, D. J., Pushor, D., & Orr, A. M. (2007). Navigating sites for narrative inquiry. *Journal of Teacher Education*, *58*, 21–35.

College of Nurses of Ontario. (2014). Competencies for entry level registered nurse practice.

Connelly, F. M., & Clandinin, D. J. (1990). Stories of experience and narrative inquiry. *Educational Researcher*, *19*(5), 2–14.

Crites, S. (1971). The narrative quality of experience. *Journal of the American Academy of Religion*, *39*(3), 391–411.

Cuevas, N. M., & Feit, M. D. (2011). *Curriculum mapping: A roadmap for curriculum coherence and student achievement*. Southern Connecticut State University.

Dennen, V., & Burner, K. (2002). The cognitive apprenticeship model in educational practice. In J. M. Spector, D. M. Merril, J. van Merrianboer, & M. Driscoll (Eds.), *Handbook of research in educational communications and technology* (3rd ed.). New York, NY: Springer.

Grol, R. P. T. M., Bosch, M. C., Hulscher, M. E. J. I., Eccles, M. P., & Wensing, M. (2007). Planning and studying improvements in patient care: The use of theoretical perspectives. *Millbank Quarterly*, *85*(1), 93–138.

Johns, C. (1995). Framing learning though reflection within Carper's fundamental ways of knowing in nursing. *Advanced Nursing Practice*, *22*, 226–234.

Knowles, M. (1984). *Andragogy in action: Applying modern principles of adult learning*. The Jossey-Bass Higher Education Series. Malco: Author.

Lewis, A. P., & Bolden, K. J. (1989). General practitioners and their learning styles. *Journal of the Royal College of General Practitioners*, *39*, 187–199.

Lindsay, G. M. (2006). Experiencing nursing education research: Narrative inquiry and interpretive phenomenology. *Nurse Researcher*, *13*, 4.

Lindsay, G. M., & Smith, F. (2003). Narrative inquiry in a nursing practicum. *Nursing Inquiry*, *10*(2), 121–129.

Magnussen, L., Ishida, D., & Itano, J. (2000). The impact of the use of inquiry-based learning as a teaching methodology on the development of critical thinking. *Journal of Nursing Education*, *39*(8), 360–364.

Mishler, E. G. (1990). Validation in inquiry-guided research: The role of exemplars in narrative studies. *Harvard Educational Review*, *60*(4), 415–442.

National League for Nursing. (2003). Position statement: Innovation in nursing education. A call to reform. Authors.

Nipissing University. (2014). *Bachelor of science in nursing scholar-practitioner program student handbook*. Author.

Polkinghorne, D. (1988). *Narrative knowing and the human sciences*. New York, NY: State University of New York Press.

Schon, D. A. (1983). *The reflective practitioner: How professionals think in action*. New York, NY: Basic Books.

Schwab, J. J. (1973). The practical 3: Translation into curriculum. *School Review*, *81*, 501–522.

Stalmeijer, R., Dolmans, D., Wolfhagan, I., & Scherpbier, S. (2009). Cognitive apprenticeship in clinical practice: Can it stimulate learning in the opinion of students? *Advancements in Health Science Education*, *14*, 535–546.

Stewart, G. W., Tondora, J., & Hodge, M. A. (2004). Evidence based teaching practice: Implications for behavioural health. *Administration and Policy in Mental Health and Mental Health Services Research*, *32*(2), 107–130.

Swenson, M. M., & Sims, S. L. (2000). Toward a narrative centered curriculum for nurse practitioners. *Journal of Nursing Education*, *39*(3), 109.

Wainwright, P. (2000). Towards an aesthetics of nursing. *Journal of Advanced Nursing*, *32*(3), 750–756.

MODELING SCIENTIFIC INQUIRY TO GUIDE STUDENTS IN THE PRACTICES OF SCIENCE: THE ED^{3}U TEACHING MODEL OF CONCEPTUAL CHANGE IN ACTION

Richard E. Shope and William F. McComas

ABSTRACT

In this chapter, we discuss ways to create optimal conditions that enable K-12 learners to experience and practice science by engaging in inquiry as an essential practice in EcoVoices, an urban environmental education program based in Los Angeles. We do this through the synergistic contributions of McComas, a science education theorist who developed the ED^{3}U instructional model; and Shope, a science education innovator who has applied, modified, and validated the model in real-world learning settings. The essential premise behind this shared work is that the more opportunity students have to practice doing science, the more authentic

Inquiry-Based Learning for Science, Technology, Engineering, and Math (STEM) Programs: A Conceptual and Practical Resource for Educators
Innovations in Higher Education Teaching and Learning, Volume 4, 217–240

ISSN: 2055-3641/doi:10.1108/S2055-364120150000004013

their science learning experience will be. And inquiry, as a science process, is one of the most important elements in the practice of science.

Welcome, students, to EcoVoices, a program of environmental discovery! Imagine all the voices you would hear if Nature could speak directly to you. EcoVoices is not your typical field trip, but an inquiry science expedition! Today, you are a member of the science community, working as scientists and colleagues to experience the thrill of discovery – especially to discover the science capabilities within yourself! (From expedition materials provided to student-scientists in the EcoVoices program)

INTRODUCTION

The vision of inquiry successfully applied in EcoVoices, but applicable much more widely in science instruction, is based on a unique and effective conceptual change inquiry instructional model called ED^3U (pronounced "ed-you") that has been used as the foundation for the ED^3U Science Performance Model. This model places the focus on students by starting with students' questions, helping them craft possible answers while providing opportunities for collection of evidence to indicate the validity of the proposed answers (Yager, 2012). This strategy aligns with the Next Generation Science Standards, which state that "Engaging in the practices of science helps students understand how scientific knowledge develops; such direct involvement gives them an appreciation of the wide range of approaches that are used to investigate, model, and explain the world" (National Research Council [NRC], 2012, pp. 42–43). ED^3U Science, as this model is often called, emphasizes multiple entryways into the iterative and interactive aspects of authentic scientific inquiry – and goes well beyond the oversimplified five-step scientific method that is described in many textbooks.

Given the constraints of the traditional school day, we have chosen to operate in an out-of-school learning realm where we can provide extended experiences for learners to become scientists and do what scientists do in the course of a field expedition. This setting offers the best of both worlds: Students and teachers attend the program as a field trip during the school day, an informal setting where they can be formally immersed in the language, culture, and practices of field research science.

THE ED^{3}U SCIENCE PERFORMANCE MODEL IN ACTION

ED^{3}U Science provides science learning experiences that emulate authentic field research. A central feature of the model is deploying science-savvy inquiry coaches to mediate conceptual change and transform teaching practice toward student-centered instruction. By "inquiry coaching," we mean a focus on science content and "providing students with a facility in the scientific way of thinking, providing an accurate picture of the way scientific knowledge is generated and has been generated in the past, and developing in students basic scientific attitudes" (DeBoer, 1991, pp. 206–207). The model calls for qualified inquiry coaches to move students through an educative science expedition experience, coaching them to think like scientists and do what scientists do. In the field of gifted education, this has been characterized as "thinking like a disciplinarian," a notion that we support for all learners.

We have implemented the model in a variety of urban educational settings, reaching more than 10,000 students with space science content each summer since 2006. Most recently, it has been applied as the science education structure of the year-round EcoVoices program, which brings busloads of 60–75 students (K-12) for day-long field expeditions in ecological science and environmental stewardship. At the site of the San Gabriel River Discovery Center, nestled along an 11-acre stretch of riparian corridor, within the urban ecosystem of Greater Los Angeles, the EcoVoices program content is focused on the regional consequences of global ecosystem change as it plays out within the San Gabriel River watershed. This patch of land and stretch of river becomes the site for the EcoVoices field expeditions. In addition to the intensive program, we also offer traveling expeditions "on the road" to schools and recreation center sites within Los Angeles County. The EcoVoices program has operated successfully for three years; the weekly expeditions have reached more than 10,000 students, and the summer traveling expeditions more than 5,000 students.

The program is facilitated by a staff of over a dozen qualified inquiry coaches – a mix of credentialed teachers, college graduates with science degrees, and undergraduates working on science or education degrees. These inquiry coaches work with students to bring inquiry to life within the framework of the ED^{3}U Science model. Just as scientists conduct investigations in a variety of settings, we craft science learning into a field expedition environment, within which students can practice science and learn more quickly as they apply the principles of scientific inquiry as part of the fun. We provide students with science expedition journals and inquiry strategy guidelines to navigate through the expedition. In EcoVoices, student-scientists draw, write, jot down questions, observations, reflections, ideas, and thoughts, keying into the inquiry strategies that they use. Our inquiry coaches review these journal entries and assist students in the practical use of their journals to extract data and highlight personal discoveries, which are then shared with others in what we call a "performance symposium" at the close of the expedition day.

Before going into the details of the ED^{3}U Science Performance Model and how it works in practice, we offer some background to explain its theoretical basis.

CONNECTING CONSTRUCTIVISM, CONCEPTUAL CHANGE, AND INQUIRY

Science instruction should be designed such that students leave the classroom with accurate views of both scientific knowledge itself and the practices of science (NRC, 1996, 2013). As in all learning experiences, students enter the classroom with prior knowledge and thoughts about science practice drawn from their own encounters with the world. Not surprisingly, students' personal conceptions are often at odds with desired scientific understandings, thus providing a challenge to their encounters with the world (Driver, Guesne, & Tiberghien, 1985; Karplus & Thier, 1967; Mintzes, Wandersee, & Novak, 1998; NRC, 1996; Osborne & Freyberg, 1985; Project 2061, American Association for the Advancement of Science, 1993). Given that students enter with prior knowledge, a robust view of the learning process is for students to develop or "construct" an accurate view of science content and process for themselves. Science instruction is not about filling minds but rather helping learners craft and recraft personal understanding. Guided inquiry can help with both tasks. Inquiry-based science learning occurs whenever students are encouraged and shown how to generate and pursue curiosity-driven questions about an intriguing natural phenomenon. Scientific inquiry begins when students are able to move beyond their personal science conceptions to pursue evidence of contradiction or corroboration with real-world data.

CONSTRUCTIVISM AS A REFERENCE FOR INQUIRY-BASED SCIENCE TEACHING

Constructivism is a vital element in effective teaching because it makes use of the reality that learners already know things when they come to class and that teachers must determine what students already know and apply that information when designing instructional plans. This view of learning has early and deep roots. Dewey (1916) urged educators to recognize that learning always involves the learners' present understanding as a starting point and that, as such, prior student experience is restructured in the mind through the process of interacting with the teacher and other students. Teaching for conceptual change takes this into account. "Individuals approach any inquiry with a prior set of concepts. The nature of these concepts significantly determines what is learned and how it is learned" (Strike & Posner, 1982, p. 232). The teacher cannot simply present content as is, but must consider what concepts and thinking processes students bring to the learning environment and how students construct or transform personal knowledge into personal understanding. This is a particular challenge in science because students have experienced the natural world and have formed personal ideas about it. In addition, students have watched media accounts of science and have constructed for themselves an account of the ways in which scientists work.

Effective science teaching begins with the student. As Carey (2000, p. 19) reminds us:

> Teachers and science educators should be made aware of the important and perhaps surprising consequences of looking at the problem of science education in terms of conceptual change. For example, I have often heard teachers and science educators blame student misconceptions on faulty education at an earlier stage in the curriculum. Rather, student misconceptions are inevitable. Not having the target concepts is not an undesirable stage in students but an absolutely necessary one. Indeed, students will construct intermediate steps and misconceptions that do not conform with the views of developed science, and educators should recognize when these steps constitute progress, not problems.

When we focus students' attention on how science knows what it knows – that is, on how science works – they learn to discern the validity of knowledge claims through strategic questioning and application of inquiry processes.

SCIENCE AS INQUIRY

The National Science Education Standards (NRC, 1996) initiative focused strongly on inquiry and said, in part, that students must (1) know that inquiry is the major process used by scientists and (2) be given the opportunity to engage in inquiry. Although the earlier National Science Education Standards were not the first to endorse such a view, the prominence they gave it renewed the conversation about inquiry.

In real-world science, inquiry is ever present. Through inquiry, scientists gather information and propose hypothetical predictions, descriptions, and explanations, all while accessing and building on their prior experiences and understanding. Scientists apply inference and other scientific inquiry methods to construct science conceptions as models, conceptual maps, and other forms derived from logical and figural thinking processes. At the end of their work, and given that there is no final arbiter of truth, the best that scientists can do is to compare notes with each other in what has become known as scientific argumentation. When scientists announce significant discoveries, they communicate not only the what, displaying and interpreting the results, but also the how, explaining the design of the investigation that produced the results (Ford, 2000; McComas, 1998; Valiela, 2001).

When we speak of having students engaged in the practice of science, we must consider the degree of student involvement in the encounter with the science phenomenon. In their early development of a teaching model called the "learning cycle," Karplus and Thier (1967) discuss four degrees of involvement: (1) The minimal amount takes place when the student reads text or listens to lectures, because there is no way of gaining insight as to whether the student has the background experience and the literacy to make the words meaningful. (2) A somewhat higher degree of involvement takes place in the midst of interactive discussion, which – though limited to verbal descriptions – has the added advantage of the give-and-take of student–teacher and student–student exchange of ideas and provides the teacher some diagnostic insight. (3) A somewhat higher degree of involvement occurs with participatory demonstration, whether student led or teacher led; those who participate have a higher degree of involvement than those who watch and discuss, and student responses provide some diagnostic insight. (4) The fourth and highest degree occurs when students are directly involved in hands-on, problem-solving activities that require active physical, social, and verbal involvement; this provides the greatest potential for diagnosis by the actively involved teacher.

Therefore, the most authentic models of science instruction integrate inquiry with an understanding that knowledge growth in students proceeds from a foundation of conceptual change. We will briefly explore several of these models in a subsequent section.

ENGAGING IN CONCEPTUAL CHANGE SCIENCE TEACHING

The need to engage in teaching that causes conceptual change in students has inspired a variety of instructional approaches that share common distinguishing features, including guiding students to the question, discussing, and testing the viability of their own ideas while recognizing that "Science is never complete. There are always some unanswered questions, some unexpected phenomena" (Karplus & Thier, 1967, p. 26).

The basic learning cycle (or teaching model) has three phases, each with a variety of names. What is often called the Three-Step Model was first discussed by Chester Lawson in his book *Language, Thought, and the Human Mind* (1958) and more fully conceptualized by Robert Karplus (late 1950s to early 1960s) and by Atkin and Karplus (1962) as "guided discovery" and used in the Science Curriculum Improvement Study (SCIS) elementary science program. In 1967, Karplus and Thier first named this instructional tool the "learning cycle" and described its component phases. Through the years, the learning cycle has been reinterpreted by Barnes (1976), Driver and Bell (1986), Karplus (1977), Erickson (1979), Novick and Nussbaum (1981), Renner (1982), Lawson, Abraham, and Renner (1989), and Rowell and Dawson (1983). As a result, there is no single design for learning cycles, but that shown in Fig. 1 appears in the literature. Opportunities for diagnosis of students' science conceptions occur in all three phases, even if teachers frequently fail to engage students in that important activity.

Bybee (1997) and the Biological Sciences Curriculum Study (BSCS) devised the "5E Instructional Model," a five-step Learning Cycle composed of the three identifiable features of learning cycles, adding emphasis at the outset to engage the students and incorporating evaluation at the end. This popular model is often implemented as a way to encourage collaborative curriculum planning among teachers and to enable analysis of teaching practice and modifications to enhance teaching effectiveness. Of course, it can also be used by informed educators to determine prior understanding on the part of students (particularly as students engage and explore the phenomenon), although this does not occur frequently enough. (Fig. 2)

In 1995, McComas proposed a conceptual change model, ED^{3}U, based on prior work in learning cycle design and informed by a strong constructivist foundation (McComas, 1995). This model is much more student and inquiry centered than prior models because it makes the connection to the role of constructivism in effective science teaching explicit. For many

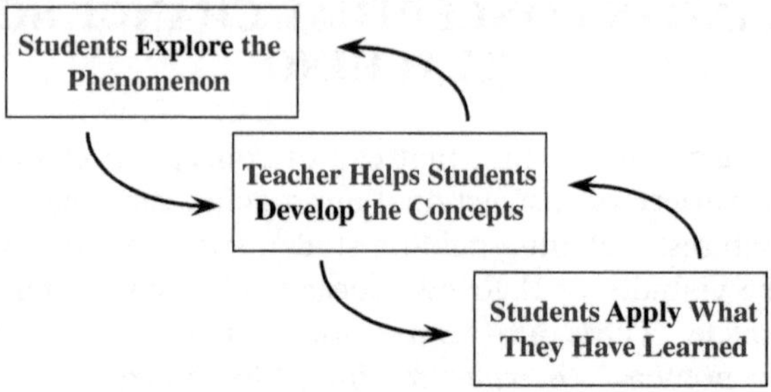

Fig. 1. Typical Phases of the Learning Cycle.

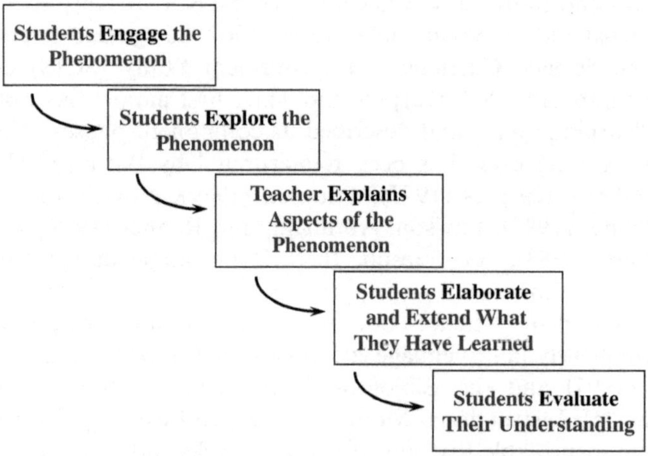

Fig. 2. The BSCS/Bybee 5E Learning Cycle Model.

teachers, this model is more difficult to use in classroom settings because the teacher's role is changed significantly. In the three-phase and 5E models, there is a time for the teacher to "label" or "explain"; in ED^3U, this "telling" (or direct instruction) is more limited and takes place primarily during discussion with students after they have explored on their own. Certainly, there are instructional situations in any science class, such as operating equipment safely, that demand a direct instructional approach. However, for those who wish to engage students in the richest of inquiry experiences, it is a logical extension to other learning cycles (Fig. 3 and Table 1).

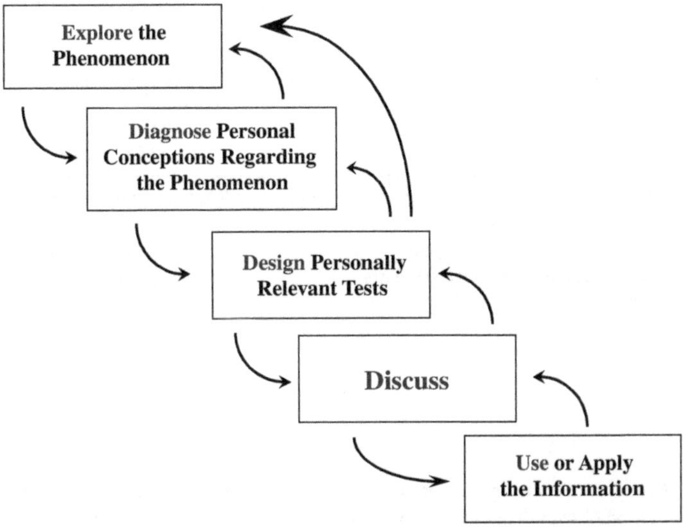

Fig. 3. The ED^{3}U Model of Conceptual Change, Based on Work by McComas (1995).

THE ED^{3}U INSTRUCTIONAL MODEL: AN OVERVIEW

The ED^{3}U Science Performance Model emphasizes the importance of diagnosing students' personal conceptions. Through diagnostic teaching strategies, the teacher – in partnership with the students – seeks to gain insight into existing conceptions and to probe the dynamics of conceptual change (Kyle, Family, & Shymansky, 1989). Diagnosing students' prior conceptions entails a degree of student "performance," a term meant to evoke images of students enacting science, applying inquiry strategies, using science instruments, participating as a community of science colleagues, and communicating science discoveries through means expected of science research professionals: writing papers, developing posters, giving presentations, attending conferences, and arguing meaning with colleagues at meetings and in symposia. Student performance generates diagnostic information. In addition, the model makes use of the idea of social constructivism as a kind of environment in which knowledge is socially mediated, negotiated, and ultimately constructed by the individual (Vygotsky, 1934/1987). The ED^{3}U model has formed the basis of student engagement in the EcoVoices and previous such programs, helping to validate, refine, and further develop it during two decades of application.

Table 1. A Comparison of the 5E and ED^3U Instructional Models.

The 5E and ED^3U Science Learning Cycle Models Compared			
Phase	5E Descriptors Bybee (1997)	Phase	ED3U Science Descriptors (McComas, 1995; Shope & Chapman, 2001)
Engage	Make connections between past and present learning experiences, and anticipate activities and organize students' thinking toward the learning of current activities.	Explore	Students explore the phenomenon: may be introduced by the teacher in the form of a discrepant event, a visual display, a hands-on activity, an observation, or exposure to a variety of information sources, print, videos, film clips, internet sites. Object is to think toward an explanation of the phenomenon.
Explore	Provide students with a common base of experiences within which concepts, processes, and skills are identified and developed.	Diagnose	Perceive student conceptions; assess how students see alternative conceptions as plausible; prescribe process that brings misconceptions to student awareness; guide student to search for a more accurate explanation.
Explain	Focus attention on particular aspects of engagement and exploration. Provide opportunities to demonstrate conceptual understanding, process skills, or behaviors. Introduce new concepts, processes, or skills.	Design	Students design personally relevant tests; create a context for a crucial experiment or prediction to test strength or weakness of a proposed explanation in both its explanatory and predictive value. The process of thinking out a personally relevant test allows the students to explore the phenomenon in some new ways.
Elaborate	Challenge and extend students' conceptual understanding and skills. Through new experiences, students develop deeper and broader understanding, more information, and adequate skills.	Discuss	Students discuss the implications. If the results suggest that a new conception is needed to replace a misconception, such alternative ideas are considered. Students may come up with their own ideas or their readiness may be open to exposure to a new theory presented by the teacher or other information source.

Table 1. *(Continued)*

The 5E and ED^3U Science Learning Cycle Models Compared

Phase	5E Descriptors Bybee (1997)	Phase	ED3U Science Descriptors (McComas, 1995; Shope & Chapman, 2001)
Evaluate	Encourage students to assess their understanding and abilities and provide opportunities for teachers to evaluate student progress.	Use	Students apply new understanding; place the new conception in relation to other related knowledge; may also lead to a new cycle that further confirms or disconfirms the validity of the new conception.

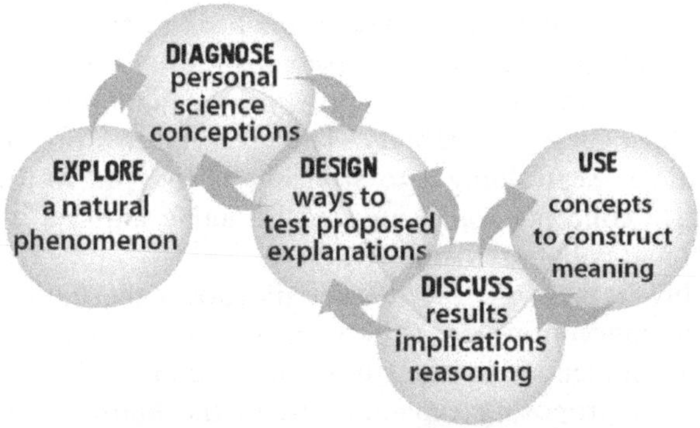

Fig. 4. The ED^3U Instructional Model as Applied in the EcoVoices Environmental Education Program.

Fig. 4 provides an overview of the ED^3U (Explore + Diagnose + Design + Discuss + Use) instructional model. The five letters refer to the phases through which the teacher diagnoses students' conceptions and/or assists students in recognizing their own notions, guides further exploration, mentors individual progress, challenges ideas, and scaffolds instruction for students. Note that the various feedback loops – indicating the kinds of actions that scientists themselves engage in – are also applied by students when learning science.

"E" signifies EXPLORE, to explore a natural phenomenon, as close to a hands-on and real-time encounter as possible. "D3" refers to a trio of related actions: (1) DIAGNOSE – to discern the nature of students' personal science conceptions. Diagnosis is vital to both teachers and students because both must understand what students think as they are thinking. The teacher can then engage students by using instructional strategies from the teacher's repertoire. (2) DESIGN – a phase in which students devise ways to carry out investigations to test personally relevant questions. And (3) DISCUSS – to generate ideas, ask questions, propose alternative explanations, make knowledge claims, reflect on implications, and interpret results; to participate in the discourse of science. "U" signifies USE – to apply and/or communicate findings, to apply newly produced scientific knowledge, to generate new expeditions.

The premise of the ED^3U Science model is conceptual change mediated and facilitated by teachers working with students to guide learners from a loosely organized system of personal conceptions toward a more informed and coherent system of conceptions consonant with the world of science. Teachers heighten students' awareness of the need to examine assumptions, as students initially draw upon a diverse substrate of their own personally relevant ideas and, through discussion, propose alternative explanations and then test their ideas for viability. The teacher helps provide assistance (sometimes called "scaffolding") to students as they move toward understanding science concepts while engaged in active inquiry, working both independently and collaboratively.

Utilizing this framework, the teacher diagnoses the nature of the students' personal conceptions throughout the lesson: in the midst of exploration of the phenomenon; in the midst of generating questions, making speculations, and proposing explanations; in the midst of designing and carrying out investigations; and in the midst of participatory discourse that actively involves students. Diagnosis is an ongoing process, monitoring the potential for conceptual change and triggering real-time shifts in instructional strategies. This calls on the inquiry coach to encourage active student discourse and active listening, an effort designed to foster a true sense of operating as science colleagues.

With this in mind, the aim of ED^3U Science is to immerse students in the language and culture of science. The structure helps students emulate how scientists conduct scientific inquiry – how they think and what they do – and, especially, to understand that science is a creative enterprise of generating questions and solving problems in the context of a science expedition experience. Conceptual change is achieved by moving science learners through three differentiated zones toward an understanding of science concepts in the midst of active inquiry, both independently and collaboratively. Conceptual change is directed as an upward spiral, continuously extending the learner's zone of proximal development, as the student moves toward mature conceptualizations (Table 2).

To explain the practical use of ED^3U Science succinctly, we will illustrate the model in three zones within which students enact conceptual change scenarios, corresponding to the underlying structure of the learning cycle: Exploratory, Discovery, and Understanding. To provide more detail of what happens in each of these zones of the ED^3U Science model, we will briefly discuss each separately.

Exploratory Zone (E = Explore the Phenomenon): Inquiry-based, in which students experience the natural phenomenon, generate questions, make speculations, and propose explanations; then explore the phenomenon in a deeper way, framed by the student-generated questions and explanations in mind; the teacher diagnoses by actively listening to a range of questions, speculations, and explanations to gain insight into conceptual change potential, returning students' questions with guiding questions.

Discovery Zone (D3 = Diagnose, Design, Discuss): Inquiry-driven, in which students utilize scientific inquiry tools to figure out what is thought to be known already, generate questions, define problems, design personally relevant tests of proposed explanations, carry out investigations, interpret results, and discuss ideas and implications all along the way; in which the teacher diagnoses student progress, introduces scientific discourse, and guides the use of science journals and science instruments in the midst of direct encounters with natural phenomena.

Understanding Zone (Use): Discovery-driven, in which students construct comprehension, show evidence of conceptual understanding, put their new knowledge to use, communicate findings as discovery highlights, apply the knowledge in a new situation, and solve a new problem by

Table 2. Nine ED^3U Scenarios of Conceptual Change in Science.

	Exploratory Zone	Discovery Zone	Understanding Zone
E	Explore, guided by curiosity	Explore further, guided by a researchable question	Produce a draft science performance project
D^3	Generate questions in midst of initial encounter	Discuss and devise ways to test proposed explanations	Share and critique project, reconstruct understanding
U	Make speculations in midst of initial encounter	Carry out a designed & data-driven investigation	Produce and publish science performance project

amplifying differentiated features, or graphically reorganizing concepts into a more coherent "big picture."

THE ED^{3}U SCIENCE MODEL IN ACTION: ECOVOICES AND THE SCIENCE EXPEDITION

The basic structure of the EcoVoices program is straightforward: a day-long inquiry-driven science expedition, focused on ecological science and environmental stewardship. EcoVoices entails a nature hike, a science show, a suite of hands-on activities, a field expedition, and a culminating symposium. Fig. 5 illustrates how the ED3U Science model translates into practice to create a science learning experience structured as a guided field expedition.

A true science expedition gathers experts together in a common mission, focused on core science objectives that guide the collecting of data needed

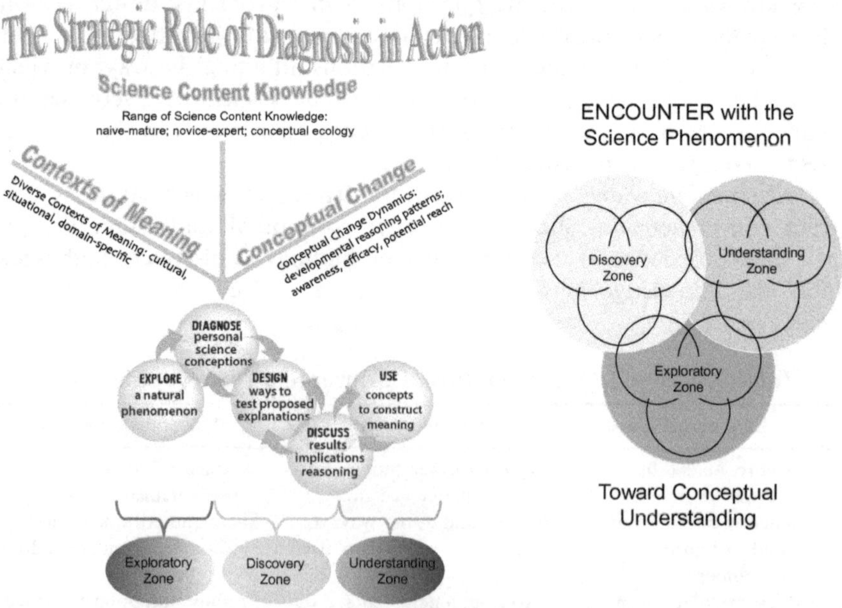

Fig. 5. Connecting the ED^{3}U Science Performance Model to the Expedition Zones.

to answer a question or solve a problem. Research is designing and carrying out the investigation. Analysis is the process of determining what the data mean, and synthesis permits construction of understanding and communicating significant results or discoveries.

Students are not science experts, but they have the potential to be motivated to pursue study and career pathways as they advance toward science expertise. The conceptual change model is designed to simulate the work of scientists while providing the thrill of discovery in the context of scientific work. Educationally, the model also accounts for the reality that learning science often requires changing students' prior conception of science and the work of science. During the expedition, students stretch their capabilities while applying scientific inquiry. In short, the ED^3U Science Performance Model emulates a science expedition within an educational context during a day-long experience.

THE FIELD EXPEDITION SITE

Typically, a field expedition site consists of the site of the featured natural phenomena, the field headquarters, and the field laboratory. At EcoVoices, the main feature is an outdoor nature center, with a small indoor natural history museum and outdoor education center, a landscaped park with picnic tables, and trails that cross into woodlands, chaparral, and access to a section of the San Gabriel River, typical of the ecosystems of a Southern California watershed. The field headquarters is a stage area setup for whole-group gatherings, performances, and symposia. The field laboratory consists of (1) a suite of "inquiry stations" where students use science instruments, examine specimens, write in science journals, design their investigations, and practice skills of participatory action research; (2) visually informative displays and exhibits in an interactive museum area, or strategically placed as portable exhibits; and (3) the field research site, out on the trails and at the river, where students have a direct, intimate experience to obtain as authentic an encounter with natural phenomena as possible.

THE EXPLORATORY ZONE

The first order is to provide a brief orientation: Students arrive and are greeted with a quick orientation. We introduce our staff of science-savvy inquiry coaches and form students into five or six research squads of 12–15 student-scientists. Each inquiry coach immediately takes their research squad of student-scientists on a 10-minute hike to explore the natural surroundings. The purpose is to activate curiosity. For example, suppose that students notice neighboring ant colonies along the trail and become curious about the territoriality of the ants. Are they getting along or are they at war? Similarly, two birds are chirping in neighboring trees. Is this friendly chatter or feuding rivalry? The role of the inquiry coach is not to answer the questions outright, but to stimulate student interest in pursuing such questions.

The Exploratory Zone is all about first impressions and establishing a basis for a discussion that gets students to talk out loud about what they are thinking during this initial encounter with natural phenomena. While the students are discussing their first impressions of the nature center site, the inquiry coach is processing the initial diagnostic impressions of the students. The nature hike is where the coach orients students and invites them to read nature and share aloud their observations – this initiates awareness of the science content. This hike, however, is NOT a typical informational tour, consisting of naming species of trees and birds and recounting natural histories; the purpose is to evoke student dialog, structured for students to listen and respond to each other. This is often counterintuitive for teachers and inquiry coaches at the outset – they may feel they are not doing their job if they are not busy explaining. In practice, as teachers and inquiry coaches succeed in evoking meaningful student participation, the purpose becomes evident: to generate diagnostic information. With such information, the coach can make real-time course adjustments in guiding the subsequent science learning experience of the student participants.

Most students will have had little experience with the specific environment of study. Therefore, the inquiry coach often stops to allow extended time for just listening, just looking, from a variety of perspectives, suggesting such things as "Get down on your hands and knees and imagine yourself as a coyote at night, or a rabbit moving through the underbrush, or think of yourself flying in the sky looking down through the eyes of the red-tailed hawk circling overhead." Exploring and observing nature involves a mix of empathy tempered with guided inquiry.

Our coaches quickly recognize that they must neither overexplain nor feign ignorance, but guide the conversation to avoid premature closure of students' thinking. Coaches work to stimulate students to ask questions, priming them not to expect immediate answers in an instant (as is often the case in school science) and to collect the most meaningful questions as candidates for further investigation, captured as notes in the science journals. In a venture into the unknown, not knowing is the starting point of scientific inquiry. If the answers were already known, if the problems were already solved, there would be no need for an expedition.

In the second part of the Exploratory Zone, we return to the field headquarters for a 15-minute whole-group briefing. We connect the students' first impressions with the theme of the EcoVoices expedition. The lead inquiry coaches illustrate examples of the target concepts of ecological science and environmental stewardship, using actual science terminology, presented for understanding as playful science banter. The briefing also incorporates discussion of scientific inquiry in reference to a larger-than-life inquiry wheel (5 feet in diameter). The banter models the discussion that student-scientists are expected to engage in as they weigh the pros and cons of each strategy in question during the expedition. Students are informed that they will also be expected to use the inquiry wheel to advance the course of their own expedition experience, which culminates with the "wrighting" aspect (i.e., writing, crafting, and rehearsing) of a performance of an EcoScenario to communicate highlights of their own expedition discoveries. For example, students may enact the movement of water through the watershed ecosystem, using mime and theatrical expression to depict evaporation, cloud formation, fire, rain, seed germination, trees growing, habitats forming, various urban uses of water, and, finally, flowing out to the ocean.

THE DISCOVERY ZONE

Inquiry coaches then take their research squads to the Discovery Zone, comprised hands-on, field-equipped inquiry stations for activities focused on concepts related to the expedition theme. Inquiry stations are designed to (1) activate curiosity; (2) evoke conceptual relationships; (3) enable acquisition of science fluency (students learn the skills they will later apply, such as when they hike to the river to retrieve water samples for quality testing); and (4) increase the probability that comprehension is constructed accurately. Each inquiry station is decked out with hands-on activity materials and equipment that is designed to trigger the formation and enhancement of key concepts.

Field equipment includes everything that a scientist would use for a field expedition. At EcoVoices, we provide (1) science journals, clipboards, and writing instruments to document expedition progress and gather data; (2) science instruments – field guides (to identify species of birds, insects, plants, trees), microscopes, binoculars, bioacoustics equipment, probes, testing equipment, specimen jars, pipettes, test tubes, hot plates, chemicals, pH paper, safety goggles, safety gloves; models, maps, magnifying lenses, graph paper, calculators, measuring tapes, and the list goes on; (3) surfaces to work on (tables, benches), seats to sit on (chairs, buckets); (4) technologies – the sound systems for the stage performances, photo and video cameras, computers or tablets; and (5) larger-than-life inquiry wheels.

The heart of each Discovery Zone activity is the student-centered creation of science enterprises. An enterprise is a student-generated project based on a problem to solve or a question to answer, of genuine interest to the students. For example, at one of the watershed science stations, students set up a series of tanks related to the stages of wastewater treatment going from the river to the tap. Students are thrown into the problem: What do we know about what goes down the kitchen sink and flushed down the toilet? What should be done with it once it all reaches the treatment plant? How could they take advantage of knowing that bacteria are already in the goop, decomposing the waste? Rather than treat the lesson as "covering the material," the inquiry coach urges the student-scientists to propose solutions, offering clues to advance the conversation at critical moments. Ultimately, the activity leads to discovering that the decomposition process can be accelerated by amplifying the existing bacteria – and that is why wastewater facilities hire microbiologists.

At EcoVoices, we set up a suite of activities related to the understanding of watershed ecosystems. For example, we set up three biodiversity stations (bird, insect, and tree communities) and three watershed science stations (water quality, water treatment, and hydrogeology). Ideally, students would work at all six stations and come away with a sense of the big picture. In practice, we have a variety of approaches. Typically, each research squad experiences one station in the morning for about 60–90 minutes, acquiring knowledge of a particular set of concepts. In the afternoon, each of the three watershed science squads switches with each of the three biodiversity squads for another 60–90 minutes. The afternoon expedition takes advantage of the newly acquired knowledge garnered from the morning expedition.

At the inquiry stations, coaches use "science banter" to evoke discussion and accommodate diagnosis. They help students set up their expedition journals. The journals are basically blank pages with a colorful EcoVoices collage on the front cover, and an inquiry wheel drawn as a compass rose on the back. The inquiry coaches start off with quick scenarios that trigger problem solving and question asking, anticipating the topic about to be presented.

For example, to understand the rate of photosynthesis of a tree, students may come to realize that they need a way to figure out the number of leaves on a tree. The inquiry coach gets a discussion going around the question "How do we go about figuring out the number of leaves on a tree?" It is up to the student-scientist research squads to come up with an efficient method to get the data – counting, estimating, sampling, extrapolating, using some algebra and geometry. An enterprise might involve proposing and trying out the ideas. Once such a method is found to work, it might be applied to other questions; for example, the leaf count might be a factor in figuring out the rate of evapotranspiration, the movement of water through a plant or tree. This information, in turn, might be useful for other questions – for example, figuring out what kind of trees might be deemed "drought-resistant."

Coaches help focus the interest of a research squad, guiding the science dialog toward framing researchable questions and explanatory hypotheses. The main job of the inquiry coach at the inquiry station is to "mediate conceptual change" from students' personal science conceptions about nature, ecosystems, and ecosystem change toward more expert understanding, coupled with more expert understanding of how science works – by doing it, reflecting upon it, and doing it some more.

One highly effective tool for creating and carrying out science enterprises – in terms of researchable questions and problem situations – is the use of the "inquiry wheel" (Fig. 6), which depicts phrases referring to scientific inquiry strategies that scientists use and have internalized in the course of their work (Reiff, 2002; Shope, 2007). For students, it can be thought of as a navigating tool, a reference guide, a conceptual compass that generates discussion and consideration of what to do next in the course of the science expedition. With practice, students become adept at arguing for the selection of an appropriate strategy, able to express the phrases in their own words and elaborate on their meaning and practicality.

The inquiry wheel is used to motivate the D^3 trio of related actions: Diagnose, Design, Discuss. The strategies depicted on the wheel help the groups engage in explicit discussion, figuring out what needs to be done and how to design the investigation. Along the way, the inquiry coaches encourage the use of the inquiry wheel as a driver of the discussion needed to conceive, design, and plan out the enterprise, focusing attention on devising a sequence of actions to carry out a field investigation. From this emerging discussion, coaches help the research squad design several possible enterprises and carry out at least one investigation in real time, in which

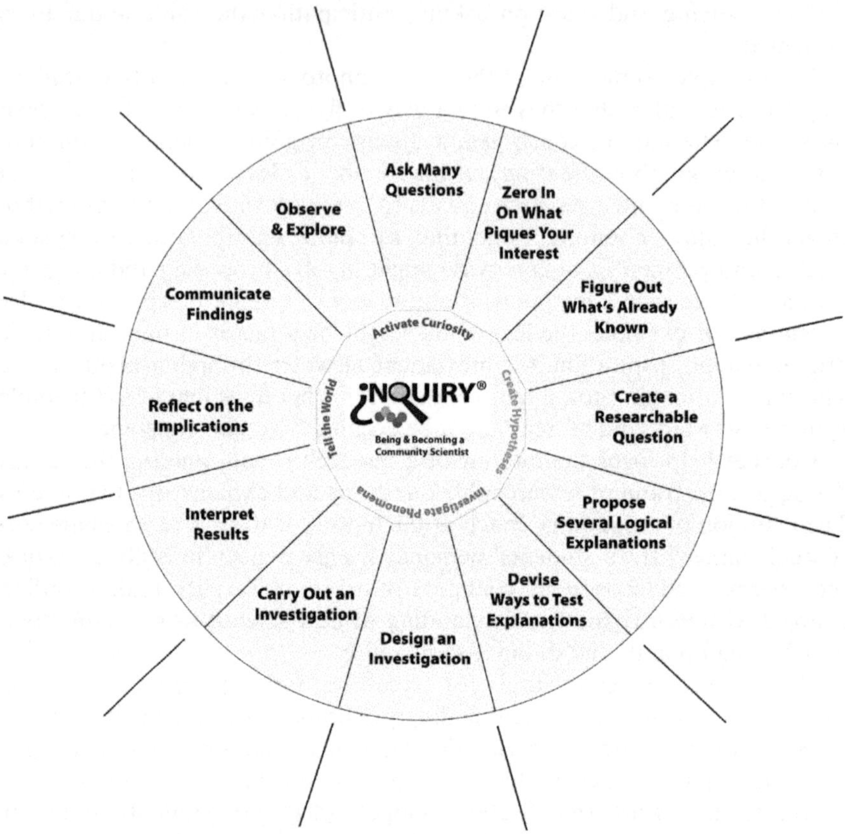

Fig. 6. The Inquiry Wheel.

students gather and organize data. They figure out what critical measurements may be required and what instruments may be needed to gather data, and then carry out as much as is actually possible within the constraints of the field expedition experience. The thinking process is then focused on interpreting and reflecting on the results.

THE UNDERSTANDING ZONE

The purpose of the Understanding Zone is for students to construct and articulate conceptual understanding of personal discoveries that emerged in the course of the expedition. In this phase, students show evidence of constructing comprehension by creating a science performance project, such as a project display, a poster, a slide presentation, a paper, a video, a lecture, or a panel discussion. At EcoVoices, we have the students work collaboratively to draw from their journals to pool their ideas. Presentations are enacted in the form of panel discussions, news flashes, or science shows using mime and improvisational theater with narration and dialog. Students write and rehearse EcoScenarios, readied for performance on stage, for review and commentary by their peers.

With help from the coaches, students prepare and rehearse to bring out their best talents for writing, narrating, public speaking, acting, mime, and improvisation – occasionally coaches step in to help direct the action for staging.

The purpose of the performance symposium is to provide a platform for students to represent new knowledge and significant results as discovery highlights, to place discoveries in relation to other related knowledge, to participate as a community of science colleagues, and to communicate science through means expected of science research professionals.

FINAL THOUGHTS: CONCEPTUAL CHANGE AND INQUIRY LEARNING

Participants arrive at the EcoVoices field expedition site at nine in the morning thinking of themselves as students on a field trip. They leave at two in the afternoon seeing themselves as a community of student-scientists, with science journals in hand as the evidence of achieved conceptual understanding. Each week, 60–75 more K-12 students have worked together as colleagues in research squads, on a science expedition driven by their own curiosity. They have moved through all three conceptual change zones, thereby completing an ED^{3}U Science learning cycle. They have explored natural phenomena in the inquiry-based Exploratory Zone. They have created their own field expedition enterprises in the inquiry-driven Discovery Zone. They have articulated their personally relevant science insights in the discovery-driven Understanding Zone. Taken as a whole, the day-long expedition provides students an authentic experience of science as inquiry and what it means to become, and to be, a true scientist.

By enacting the ED^{3}U Science Performance Model as a field expedition, we have helped thousands of K-12 students acquire the practices of scientific inquiry. We have accomplished this goal by guiding students on inquiry expeditions in which they move through conceptual change zones where they (1) experience actual practice in the role of an inquiring scientist, (2) see how scientific knowledge is generated over time, (3) apply methods of scientific inquiry, (4) read nature and write scientific communications, (5) enact science scenarios, and (6) emerge with a self-identity of becoming and being a scientist.

The EcoVoices expedition has generated a great deal of enthusiasm among participating teachers. We plan to conduct an evaluation, utilizing a science-educative-experience framework suggested by Tobin (1993, 2007), not only to validate the perceived effectiveness of the underlying ED^3U Science Performance Model, but also to examine its transferability as a means to transform teacher practice toward more authentic inquiry experiences for students during the school day.

REFERENCES

Atkin, J. M., & Karplus, R. (1962). Discovery or invention? *The Science Teacher*, *29*(5), 45–51.

Barnes, D. R. (1976). *From communication to curriculum*. New York, NY: Penguin Education.

Bybee, R. W. (1997). *Achieving scientific literacy: From purposes to practices* (pp. 178–179). Portsmouth, NH: Heinemann.

Carey, S. (2000). Science education as conceptual change. *Journal of Applied Developmental Psychology*, *21*, 13–19.

DeBoer, G. E. (1991). *A history of ideas in science education: Implications for practice* (pp. 206–207). New York, NY: Teachers College Press.

Dewey, J. (1916). *Democracy and education: An introduction to the philosophy of education*. New York, NY: Macmillan.

Driver, R., & Bell, B. (1986). Students' thinking and the learning of science: A constructivist view. *School Science Review*, *67*, 443–456.

Driver, R., Guesne, E., & Tiberghien, A. (Eds.). (1985). *Children's ideas in science*. Milton Keynes, UK: Open University Press.

Erickson, G. L. (1979). Chidren's conceptions of heat and temperature. *Science Education*, *63*(3), 323–336.

Ford, E. D. (2000). *Scientific method for ecological research*. Cambridge: Cambridge University Press.

Karplus, R. (1977). Science teaching and the development of reasoning. *Journal of Research in Science Teaching*, *14*, 169–175.

Karplus, R., & Thier, H. D. (1967). *A new look at elementary school science: Science curriculum improvement study*. Chicago, IL: Rand McNally.

Kyle, W. C., Jr., Family, E., & Shymansky, J. A. (1989). Enhancing learning through conceptual change teaching. Research Matters – to the Science Teacher No. 8902. NARST, Reston, VA. Retrieved from https://www.narst.org/publications/research/concept.cfm

Lawson, A. E., Abraham, M. R., & Renner, J. W. (1989). A theory of instruction: Using the learning cycle to teach science concepts and thinking skills. NARST Monograph No. 1.

Lawson, C. (1958). *Language, thought, and the human mind.* East Lansing, MI: Michigan State University Press.

McComas, W. F. (1995). ED^3U Model. [Notes taken by Shope in a class taught by McComas. Unpublished document, available from the authors by request.]

McComas, W. F. (1998). The principal elements of the nature of science: Dispelling the myths. In W. F. McComas (Ed.), *The nature of science in science education: Rationales and strategies.* Dordrecht: Kluwer Academic Publishers.

Mintzes, J. J., Wandersee, J. H., & Novak, J. D. (Eds.). (1998). *Teaching science for understanding: A human constructivist view.* San Diego, CA: Academic Press.

National Research Council. (1996). *National science education standards.* Washington, DC: National Academies Press.

National Research Council. (2012). *A framework for K–12 science education* (pp. 42–43). Washington, DC: National Academies Press.

National Research Council. (2013). *Next generation science standards.* Washington, DC: National Academies Press.

Novick, S., & Nussbaum, J. (1981). Pupils' understanding of the particulate nature of matter: A cross age study. *Science Education, 65*(1981), 187–196.

Osborne, R., & Freyberg, P. (Eds.). (1985). *Learning in science: The implications of children's science.* Auckland: Heinemann.

Project 2061, American Association for the Advancement of Science. (1993). *Benchmarks for science literacy.* New York, NY: Oxford University Press.

Reiff, R. (2002). Scientists' conceptions of scientific inquiry: Voices from the front. Paper presented at the 2002 National Association for Research in Science Teaching (NARST), New Orleans, LA.

Renner, J. W. (1982). The power of purpose. *Science Education, 66*(5), 709–716.

Rowell, J. A., & Dawson, C. J. (1983). Laboratory counterexamples and the growth of understanding in science. *International Journal of Science Education, 5*, 203–215.

Shope, R. E. (2007). INQUIRY coaching: Scientists and science educators energizing the next generation. Paper presented at the 2007 Fall Meeting, American Geophysical Union (AGU), San Francisco, CA.

Shope, R. E., & Chapman, L. (2001). The space exploration team inquiry model: Linking NASA to urban education initiatives. *Association for Science Teacher Educators, 2001 Proceedings.*

Strike, K. A., & Posner, G. J. (1982). Conceptual change and science teaching. *European Journal of Science Education, 4*, 231–240.

Tobin, K. (Ed.). (1993). *The practice of constructivism in science education.* Hillsdale, NJ: Lawrence Erlbaum.

Tobin, K. (2007). Structuring success in science labs. Presentation at the Association for Science Teacher Education, Clearwater, FL.

Valiela, I. (2001). *Doing science: Design, analysis, and communication of scientific research.* New York, NY: Oxford University Press.

Vygotsky, L. S. (1934/1987). Thinking and speech. In R. W. Rieber & A. S. Carton (Eds.), *The Collected Works of L. S. Vygotsky* (Vol. 1). Problems of General Psychology. New York, NY: Plenum Press.

Yager, R. E. (June 29, 2012). Issues regarding use of "inquiry" vs. "practices" for the Next Generation Science Standards (NGSS). NSTA Blog. Retrieved from http://nstacommunities.org/blog/2012/06/29/issues-regarding-use-of-inquiry-vs-practices-for-the-next-generation-science-standards-ngss

UTILISING CONTEXT IN LEARNING

Wendy Trimmer and Philip Hawes

ABSTRACT

This chapter discusses the implementation and current use of an inquiry-based approach to teaching and learning at Whitireia Community Polytechnic (Whitireia), a tertiary education institution, known for its high student success rate and culturally diverse student body. The Faculty of Health at Whitireia have developed the use of a context-based learning (CBL) approach across a range of courses in several health disciplines; these include nursing (undergraduate and postgraduate), paramedic, and foundational support work courses.

CBL is fully integrated in undergraduate nursing courses; students practice clinical skills in a range of simulated situations, and interact with virtual families through a variety of CBL scenarios. To varying degrees, CBL concepts are used widely in most other programmes within the Faculty. This chapter focuses on the development of the CBL approach at Whitireia, encompassing the simulated learning environment, and the use of the

Inquiry-Based Learning for Science, Technology, Engineering, and Math (STEM) Programs: A Conceptual and Practical Resource for Educators
Innovations in Higher Education Teaching and Learning, Volume 4, 241–256

ISSN: 2055-3641/doi:10.1108/S2055-364120150000004014

'virtual family'. Developing a CBL approach has been facilitated by the creation of a purpose-built physical environment. Implementing a CBL approach has required professional development for staff to enable them to competently teach in this paradigm.

BACKGROUND

Whitireia offers over 100 programmes of study covering certificate, diploma, degree and postgraduate levels. All programmes have been developed in conjunction with industry, to offer excellent applied qualifications to over 8,500 students (5,030 equivalent full-time students), who study at one of four contemporary campuses in New Zealand. The following Whakataukī (Māori saying) is the foundation stone of Whitireia:

> Ko te manu e kai i te miro, nōna te ngahere, ko te manu e kai i te mātauranga, nōna te ao. The one who partakes of the flora and fauna, that will be their domain. The one who engages in education, opportunities are boundless.

The Faculty of Health at Whitireia has always used constructivist pedagogy in programme delivery; the journey using an inquiry-based approach began in 2007 with the implementation of problem-based learning (PBL). Following evaluation research and initial staff development the approach changed to context-based learning (CBL) in 2009. CBL is a form of inquiry-based learning focused on the context of the situation rather than an identified problem. Evidence used to support this chapter utilises research concerning both CBL and PBL.

CONTEXT-BASED LEARNING

CBL can be defined as a learning process where the student is presented with real-life scenarios which are designed to replicate the social and political context of their working environment. Offricht (2012) states that 'Context plays an enormous part in understanding. Without context, events have little meaning' (p. 12). CBL scenarios are intentionally created in order for the students to navigate towards a successful outcome. This enables learning from real-life events and experiences, utilising reflection as a tool for change and development (Forneris & Peden-McAlpine, 2006). CBL is a student-centred 'group-based teaching technique' (Ahlfeldt,

Mehta, & Sellnow, 2005, p. 9), where the process of working together creates discourse and leads students towards a solution focused outcome.

CBL provides an alternative to the more traditional lecture approach to teaching. White, Amos, and Kouzekanani (1999) conducted two qualitative and two quantitative studies and concluded that PBL is an effective teaching strategy for nursing educators, and that this approach '... injects a renewed enthusiasm into the classroom environment'. Alexander, McDaniel, and Baldwin (2005) agree stating '... students gain a better understanding of the application of knowledge in the workplace and health care environments that place emphasis on evidence-based practice' (p. 118).

As a learning concept CBL originated in Canada in the 1960s at McMaster University as PBL and was used initially to prepare medical students for practice (Alexander et al., 2005; Dochy, Segers, Van den Bossche, & Gijbels, 2003). Since that time, PBL has spread throughout the developed world, including Canada, Australia, and the United Kingdom (Alexander et al., 2005). CBL has been a relative newcomer to nursing education, successfully used in nursing programmes at the University of Alberta, Canada (Profetto-McGrath & Worrell, 2010). CBL was embraced hesitantly at first, evaluative research about its effectiveness or usefulness to nursing and teaching is growing (Carlisle & Ibbotson, 2005; Chan, 2008). This approach is seen as a potentially transformative tool for building the capability, creativity and the skills of critical analysis in the profession (Alavi, 2005; Mc Millan et al., 2007).

Dochy et al. (2003) provide a meta-analysis of 43 empirical studies of CBL in tertiary institutions, and whilst it is acknowledged that there is slightly less knowledge initially acquired in CBL than in a 'traditional curriculum', Dochy et al. found that students retain more of the knowledge they acquire with a CBL teaching method. It is the application of knowledge that students retain in the applied discipline of nursing that is seen as building capability and confidence in professional practice. Chaves, Baker, Chaves, and Fisher (2006) acknowledge that the role of the educator can be challenging, 'Tutors carry the responsibility to model and nurture a complex concept of professionalism, which includes commitment to self-assessment and life-long learning as a basis for improved performance, and peer review as a basis for quality professional practice' (p. 30).

From the literature it is evident that students participate more when steps are taken to actively engage them (Ahlfeldt et al., 2005; Alavi, 2005; Prince & Felder, 2007). The goals of engaging and meeting the changing needs of students are a constant in education. Research undertaken with university students in the United States of America revealed that levels of engagement were higher in classrooms that utilised an inquiry-based approach rather than the traditional didactic approach (Ahlfeldt et al., 2005). CBL as an inductive method of teaching and learning includes trial and error in the classroom, while group discussion is seen as powerful (Prince & Felder, 2007). CBL enables students to critique and challenge social and political issues, and this process '... demonstrates critical engagement with the experiences of clients; recognition of common issues across individual client experience; the ability to synthesise, analyse, problem-solve, and reason; and creative and compassionate thinking' (Cooke & Matarasso, 2005, p. 246), all of which is essential in the current health care environment.

A CBL approach promotes '... self-directed learning, critical analysis, clinical reasoning and problem-solving skills' (Cooke & Matarasso, 2005, p. 244). Chaves et al. (2006) add to this list of benefits by stating that this approach 'requires skills in seeking and synthesising knowledge, communicating verbally and nonverbally, and engaging in interpersonal relations' (p. 30). Farideh Esfandiary's (1995) dissertation qualifies the findings that showed an increase in critical thinking and problem-solving skills in nurse practitioners who participated in CBL, showing that '... interactions were of central importance in the development of critical thinking' (p. 88). Fenwick's (2002) study of mid-career professionals' experiences and satisfaction levels of an inquiry-based approach suggest that it is very useful to long-term learning. Managing group process, understanding different perspectives, systems thinking and self-knowledge were ranked as benefits of this approach by participants.

DEVELOPING A CBL APPROACH AT WHITIREIA

The Faculty of Health, Whitireia New Zealand, a tertiary education provider, is a national leader in professional and vocational education for the health industry, ranging from foundation to postgraduate level. Programmes provide equitable opportunities for development of skill enhancement, leadership capacity and expertise, and culturally safe practice. Whitireia supports culturally diverse perspectives of health to reflect the population of Aotearoa New Zealand; for example, three undergraduate nursing programmes (Bachelor of Nursing; Bachelor of Nursing Pacific; and Bachelor of Nursing Māori) contribute graduate nurses to the health workforce who are able to work in both mainstream and cultural paradigms. Whitireia delivers a one-year full-time foundation education programme, in either health science or social science that assists students to develop skills and knowledge to enable them to staircase into further tertiary study. In addition, Whitireia delivers a Bachelor of Health Science (Paramedic) and Bachelor of Social Work. The Faculty of Health is the second largest provider of postgraduate nursing education in New Zealand and one of only two providers of paramedic education in New Zealand.

In 2007, PBL was implemented into a postgraduate nursing programme. This programme is delivered over a year with students (predominantly new graduate nurses) attending five weeks of theory, with the remainder of the learning occurring in the clinical setting. Student evaluations from 2006 indicated that content delivered was of a similar nature to their undergraduate programme, with students identifying that they wanted more guest speakers, and some students struggling to see how particular information was relevant to them and their clinical practice. During 2006, the programme's external academic advisor, Dr Christine Alavi, a specialist in PBL (having used it successfully in undergraduate nursing education in Australia) delivered a PBL workshop to a small number of staff. This workshop, along with the student evaluations provided the impetus to implement a PBL approach.

Preparation began with introducing programme staff to the idea of using PBL, one staff member felt like they were already doing it, while the other staff members were receptive to the idea. Work began by reviewing the curriculum. Excitement grew as staff began to develop scenarios that were based on the learning outcomes and content. A scenario based on the patient 'Jason Hyslop' was developed that encompassed content such as depression, suicide risk, psychopharmacology, mental health legislation, grief, family and support, recovery, and risk assessment. Along with the brief scenario that would be given to students, background information was developed detailing Jason's life story. Given the importance of making the scenario realistic, it was then sent to an expert in the clinical area, one of the programmes' stakeholders, to attain validation and check for veracity. Through this process, the scenario crafting was refined.

The next challenge centred on changing the style of teaching, as no longer were staff providing compartmentalised lectures to students, rather the role was of facilitators in a student-driven process. This took preparation and the creation of a phrase that is still used today within the faculty when discussing this approach to teaching 'Trust the Process!' The process used was one where the students are given a scenario; they identify situations in need of improvement, hypothesise, question, accept, reject or hold their hypothesis and then identify learning needs. Students then research and present their findings, make a clinical judgement, decide on actions and then reflect and evaluate depending on the outcomes of their interventions. Often during that first PBL session, staff had to 'hold their tongue' as the temptation of overtaking the process and reverting back to didactic teaching was strong. When staff did 'trust the process', they found that students discovered and presented the information that staff wanted to tell the students themselves.

In preparation for the first CBL session, students were given information about the CBL process and what to expect from the experience. Some students relished the different approach to teaching, while others felt challenged by it. When students were given the scenario and asked to identify situations in need of improvement, they immediately reverted to problem solving, sharing how they would manage Jason's care. When asked to hypothesise about the scenario staff and students could feel a transformation within the room, students were encouraged to think 'outside of the square' in relation to the scenario. Even during the morning tea and lunch breaks, students continued to talk about Jason and what might be going on for him. The energy was contagious; staff began to actually believe that this process was going to work. During that first PBL session some students did relay their anxiety about presenting in front of the class, and some felt challenged with making judgements in relation to Jason's scenario.

We came to understand how valuable this way of teaching and learning would be, not only were students obtaining the required knowledge, it appeared to us that they were getting the added benefits of learning how to apply their knowledge and evidence-based practice to the situation. Students would work with their peers and gain the skills of working in teams, which is an extremely important component of clinical practice. Students would build their confidence in relation to presenting and giving their opinion (the clinical judgement). Students would have to look beyond the initial presentation of their client/patient to be seen as safe skilled practitioners. By co-teaching, students could see that staff could have different opinions and still effectively work together, this staff discourse mirrored clinical practice. What was once an exhausting week of theory with the students turned into a week of surprise, energy, fun hard work and challenges. Other faculty staff noticed the team's sustained enthusiasm for teaching and learning.

Following evaluative research and feedback from staff, PBL was replaced with CBL in 2008, driven by staff's recognition of the importance of context and the social political environment. A CBL approach has been used consistently since that time in both undergraduate and postgraduate nursing programmes, with the Bachelor of Nursing programme staff having developed a range of CBL scenarios that are integrated throughout all papers of the curriculum. Since 2010, CBL has been utilised further within the faculty, with the Bachelor of Health Science (Paramedic) and the National Certificate in Mental Health and Addiction Support programmes implementing CBL concepts into the teaching practice.

The CBL process typically commences with a scenario, with students noticing, questioning and recognising their own learning needs. Students then accept the challenge of finding best practice (evidence) in small groups, and present this as a discussion within the wider group, applying learning to practice and making clinical judgements. Educators are supported to build their own capability in moving from a didactic approach to facilitating sessions where they make a conscious effort to question each other, create discourse, whilst mirroring a professional approach to conflict management.

Since its introduction, CBL at Whitireia has grown in popularity as a student-centred method of teaching and learning. The implementation of CBL was a purposeful strategy designed to engage learners in developing clinical and critical thinking. The CBL approach utilised at Whitireia creates an environment and level of engagement for the student not only to learn the knowledge and skills required but also to develop the essential skills of communication, team work, engagement and confidence in reasoning (Trimmer, Laracy, & Love-Gray, 2009). The rapidly changing environment in health care throughout New Zealand and across the world means that the use of the social and political contexts in crafted scenarios better prepares the undergraduate for the working environment (Kantar, 2014). The CBL process utilised at Whitireia engages the students in life-like simulations where they can apply learned theory in a contextual framework, creating a more resilient, work ready, and safer practitioner.

CBL also creates the space to close the theory practice gap for students through authentic learning experiences (Trimmer et al., 2009). Clinical scenarios are crafted using expert clinicians to ensure relevance and realism in the classroom. The CBL process allows students to think critically about practice issues, challenge ideas, use others as a resource, and ultimately decide how they would apply the evidence to their practice. In student evaluations of the programme, over 80% indicated that CBL 'contributed a great deal' to their ability to see the bigger picture of health care (Trimmer et al., 2009).

Clinical judgement is an essential skill for health care professionals, as it is at the core of effective therapeutic care. Whitireia has incorporated the Oregon Consortium for Nursing Education's model into the curriculum, which uses Tanner's Clinical Judgement Model (2006) as a base. This model emphasises the use of simulation as an integral learning tool in developing clinical judgement skills. Clinical judgement is a highly complex skill, which is nurse and context dependent, and requires critical, creative thinking and decision making (Tanner, 2006). Tanner's research-based model of clinical judgement has four components, noticing, interpreting, responding and reflecting. Clinical judgement is the best conclusion a health professional can reach at any point in time, given the information available to them (Gillespie & Peterson, 2009). Tanner's model mirrors the process of CBL, where students identify their initial response, consider their clinical skills and learning needs, interpret and respond to the scenario, as well as reflect on the effectiveness of treatment given. The incorporation of Tanner's (2006) Clinical Judgement Model into the CBL process has greatly enhanced the focus and benefit of using CBL. This combination is used in both nursing and paramedic programmes within the faculty.

A Simulated Learning Environment

In various courses, students practice clinical skills in a range of simulated situations generated through CBL scenarios. This provides students with the opportunity to navigate practical, authentic clinical scenarios, while perfecting techniques and skills relevant to their chosen career path. Simulation scenarios at Whitireia are intentionally developed and crafted to maximise learning experiences within the CBL paradigm.

Simulation addresses the problem, recognised throughout the world, that clinical placements are becoming harder to access for health care students and in particular for nursing students. The amount of clinical experience that undergraduates are being exposed to has raised fears that graduates are underprepared and potentially unsafe to enter the workplace they are being trained for (Jefferies, 2007). In response to this problem, many institutions have turned to simulation as a possible solution, as an alternative form of learning experience. Simulation is a tool that simulates real-life situations using various mediums and offers the student an immersive experience, of being placed in a 'real-life' environment. The participants are then required to use critical thinking and problem-solving skills in order to achieve a positive outcome (Lateef, 2010). Scenarios can be run in 'real time' to add further authenticity to the experience.

Simulation is more effective when used in small groups; this was a challenge to certain programmes such as the Bachelor of Nursing, as they had up to 80 students in the final year. A radical solution was proposed where students would be taken out of clinical placements in groups of eight for one day per term and given intensive, simulated, CBL scenarios. This did not impact on the mandatory required clinical practice hours required by the New Zealand Nursing Council. These intensive CBL days focused on situations that students may have not encountered in their clinical placements. Assessment, communication and leadership skills were all tested in the simulated learning environment (SLE) and feedback from students reported that they felt an increase in confidence after attending the days, and felt more able to deal with the rigours and stresses of the clinical environment (McCaughey & Traynor, 2010).

The Faculty of Health's four purpose-built simulation laboratories offer both low and high fidelity experiences for the undergraduates. Low fidelity simulation is centred on skill acquisition and is used for students who are at the beginning of their learning journey, where they develop the motor skills that are pertinent to their chosen profession. High fidelity is for the more senior students and is often used in assessing the ability of the student to be considered work ready, in line with Tanner's Clinical Judgement Model (2006). High fidelity simulation uses sophisticated manikins who have the ability to talk and interact with the participants, have vital signs, for example, blood pressure and pulses, as well as breathing and heart sounds. In simulation, the use of actors playing distressed relatives and other patients add authenticity to the scenarios. Tutors facilitate the scenarios from a control room, which has one way mirrors, where there is capacity to record the scenario in order to play back to the participants; this assists in the debriefing process. Debriefing takes more time than the scenario itself, it is essential that participants are allowed to explore and unravel the events in a positive reflective forum.

BL scenarios are developed as part of a wider package so that they fit into the context of the broader subjects that are being taught. Once verified for accuracy, these scenarios are enhanced by the use of props such as medical notes and X-rays that add to the authenticity of the experience. The more realistic the scenario, the more likely the students will become immersed in the process. Within this approach, is the notion of the fiction contract, this concept maintains that what makes a scenario successful is not just how realistic it appears, but how the participants and tutors interact within the scenario. A fully immersive experience entails emotional engagement as well, as a physical one, this engagement is not just the responsibility of the participants but also the educators (Rudolph, Simon, & Raemer, 2007).

The constructivist epistemology that underpins the Faculty of Health at Whitireia supports the student-centred CBL approach that incorporates simulation and virtual families. Kantar (2014) asserts that a constructivist approach is considered to be essential in order to prepare graduates for the realities of their practice, and states 'Construction begins with engaging learners in the meaning-making process and ends with enabling them to handle problems of the real world' (p. 234). Research undertaken at Whitireia (Trimmer et al., 2009) found that utilising constructivist pedagogy built capacity in both learners and teachers. Students were challenged to move away from the 'quick fix' of problem solving using clinical scenarios, and utilising constructivist pedagogy, and they began thinking critically, generating new knowledge.

INTERACTING WITH VIRTUAL FAMILIES

Using the synergy of simulation and CBL, the concept of a virtual family was developed at Whitireia. In undergraduate nursing courses, students meet and interact with virtual families through different CBL scenarios. This virtual family is an online resource that can be accessed by students via the Moodle online learning platform in order to familiarise themselves with the different members of the family and prepare them for teaching sessions using simulated scenarios. Created and developed by the teaching staff, the virtual family reflects the local community, both ethnically and socio-politically. This family consists of five generations covering the lifespan of new born to older adult, some family members have existing health conditions, which may deteriorate over the three years the students are exposed to them. Other members of the virtual family are healthy, but may develop health concerns during the progression of the undergraduate programme.

The virtual family is used to enhance learning in topics such as research, communication, leadership, assessment, as well as resource and crisis management (refer to Fig. 1). The rationale for developing the virtual family was to add realism and a sense of connection that the students could make with the characters. Sometimes actors play the family members in specific

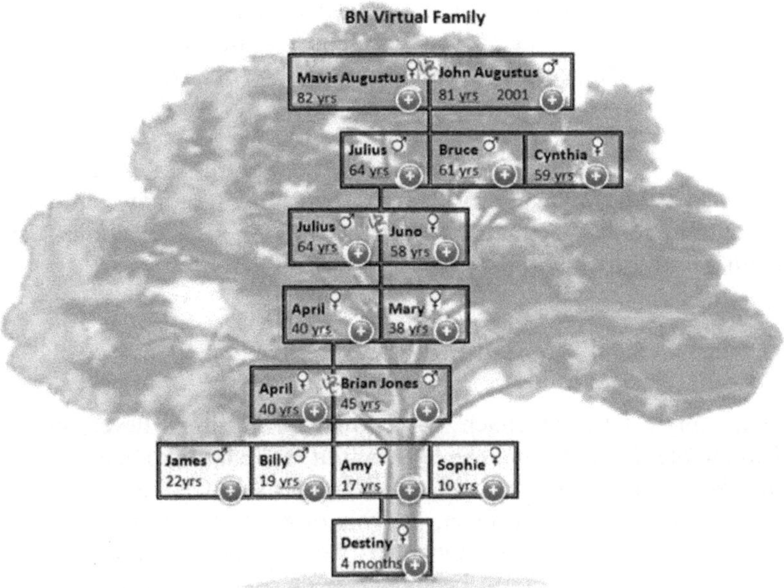

Fig. 1. Bachelor of Nursing Virtual Family.

scenarios, which gives the students a rich narrative on to which to hang their learning and enables them to emotionally engage with the members of the virtual family. This engagement results in a deeper and more profound learning experience, as found by Bizzocchi and Schell (2009). Students work individually and in groups on the various clinically relevant scenarios, in a programme where the learning is scaffolded, increasing in complexity over time.

An example of this can be found with the use of Julius Augustus, the patriarch of the virtual family. Julius is an elderly gentleman who suffers from several illnesses including chronic obstructive airway disease, type 2 diabetes and gout. A whole day is developed where students encounter Julius in the simulation suites. This day is developed in order to measure the student's assessment skills, utilising Tanners Clinical Judgement Model. The students are faced with three scenarios, centred on Julius, where they are required to go through Tanner's process of noticing, interpreting, acting and reflecting. Students work through each scenario in small teams, interpreting Julius' needs and condition, then making the appropriate clinical interventions. While all these assessments are being undertaken, an actor is used to play another patient or relative. The actor acts as a distraction to the students, getting more agitated during the scenario. The purpose of the actor is to challenge the student's ability to cope with the distraction. The students are then debriefed and are encouraged to explore feelings around performance and how they might approach a similar situation in clinical practice. Julius Augustus and his 'virtual family' becomes a familiar scenario as students' progress though their undergraduate programme, understanding Julius in the context of his family, his community, and in different health care environments.

Professional Development for Staff

Implementing a CBL approach has required professional development for staff to enable them to competently teach in this paradigm, supporting the change from didactic to a student-centred approach. CBL is different from the traditional didactic approach; the role of academic staff is to facilitate the learning process, assisting students to be creative, critical thinkers in an open environment. Lecturers become a resource for students '... rather than being a purveyor of information' (Alavi, 2005, p. 5). To ensure an effective learning experience for the students, the Faculty teaching staff are supported to understand and acknowledge the benefits of CBL strategies and simulation, through professional development and mentorship. Some resistance to the introduction of CBL was encountered from staff, this mixed response reflecting educators' natural reluctance to change practice (Alexander et al., 2005).

To ensure the successful integration of CBL across the Faculty, a needs analysis was undertaken to identify what was required in order to support skill development in facilitation of CBL. From the results of this analysis, an educational plan was developed and implemented based upon the 'All, Many, Few Model' (Holloway, 2011). This model identifies the core learning experiences that are required for all staff, then to the many staff using CBL regularly and finally to the few staff who become experts and champions in CBL. These few staff act as repositories of knowledge that all staff can access when necessary.

Professional development focused on CBL has occurred with faculty staff on a group and an individual basis to support and integrate the approach within the classroom. Professional development workshops demonstrating the nature of simulation, use of the virtual families and the benefits of these resources in CBL have been undertaken across the faculty.

This also involved faculty staff developing, participating and debriefing a CBL scenario to understand what the students experience entails, and the benefits that can be taken from participation. An understanding of the importance and subtleties of debriefing was also vital for the development of the teaching staff. Professional development was facilitated by existing CBL champions' within the Faculty of Health.

Staff development CBL workshops have also covered areas such as scenario crafting, managing resistance and the use of facilitation and integration into the curriculum. The workshops allowed for the development of scenarios, facilitation skills and team building. It also gave the participants enthusiasm and confidence to introduce CBL into their programme. The workshops were well received by participants and allowed them the opportunity to introduce CBL into their own papers, which they did in teaching teams, with support and mentoring from identified champions. The CBL process was role played by facilitators and staff, enabling the staff to acquire a practical understanding of the concepts.

CBL workshops were initially undertaken with undergraduate nursing and health programmes staff as part of a flexible learning project. The LATAR (learning outcomes, assessments, topics, activities, and resources) framework (Haggerty, 2009) was developed as part of this project, and is utilised across the faculty. LATAR incorporates the Seven Principles of Good Practice (Chickering & Gamson, 1987) and as an important instrument for curriculum alignment, development of CBL packages, and lesson planning.

Creating a Suitable Physical Environment

Recognising the contribution that CBL has made to teaching and learning, Whitireia has responded by creating a suitable physical environment conducive to CBL concepts and curriculum. The new Faculty of Health building, named Wikitoria Katene, incorporates innovative designs facilitating health education, research, and practice. This building includes learning streets to create connectedness, a state-of-the-art simulation suite, and classrooms with moveable walls to reconfigure teaching spaces easily. The environment challenges educators to be innovative in order to facilitate the learning process.

The purpose-built CBL-shared learning spaces (learning streets) contain seating, tables, access to Wi-Fi and large screens to encourage information sharing, research skills and group participation. These environments encourage students to gather information, participate in discussion and share findings. Students then reflect on experiences for the purposes of analysis and ongoing action, this creates an opportunity for self and peer assessment, outside of the classroom environment.

Excellent library facilities, with computers and internet access, academic learning support services and peer tutoring are conducive to supporting effective use of CBL. In addition, the library and IT staff at Whitireia have worked to implement a new 'student commons' space to accommodate a CBL approach. Academic Support Staff have embraced this approach to learning particularly in the first year of the Bachelor of Nursing programme where co-teaching of CBL scenarios occur that have a focus on academic writing.

CONCLUSION

The integration of CBL across the faculty has promoted the sharing of teaching resources among the differing programmes, with educators sharing expertise. The constructivist epistemology that underpins the Faculty of Health lends itself well to the student-centred CBL approach that incorporates classroom scenarios, simulation, virtual families and staff development. CBL has added to creativity within the classroom environment and promotion of critical thinking and group work skills within the student population. Ongoing research is being undertaken in order to capture CBL in action at Whitireia. The purpose of the research is to further evaluate the integration of CBL within the Faculty of Health, aiming in particular to increase our understanding of the impact of using CBL on staff and students.

REFERENCES

Ahlfeldt, S., Mehta, S., & Sellnow, T. (2005). Measurement and analysis of student engagement in university classes where varying levels of PBL methods or instruction are in use. *Higher Education Research and Development*, *24*(1), 5–20.

Alavi, C. (2005). *Problem-based learning in a health sciences curriculum*. New York, NY: Routledge.

Alexander, J. G., McDaniel, G. S., & Baldwin, M. S. (2005). If we teach them to fish: Solving real nursing problems through problem-based learning. *Annual Review of Nursing Education*, *3*, 109–123.

Bizzocchi, J., & Schell, R. (2009). Rich-narrative case study for online PBL in medical education. *Academy of Medicine*, *84*(10), 1412–1418. doi:10.1097/ACM.0b013e3181b6ead0

Carlisle, C., & Ibbotson, T. (2005). Introducing problem-based learning into research methods teaching: Student and facilitator evaluation. *Nurse Education Today*, *25*(7), 527–541.

Chan, L. C. (2008). The role of a PBL tutor: A personal perspective. *Kaohsiung Journal of Medical Science*, *24*(3), 34–38.

Chaves, J. F., Baker, C. M., Chaves, J. A., & Fisher, M. L. (2006). Self, peer and tutor assessments of MSN competencies using the PBL-evaluator. *Journal of Nursing Education*, *45*(1), 25–31.

Chickering, A., & Gamson, Z. (1987). Seven principles for good practice in undergraduate education. *American Association of Higher Education Bulletin*, *39*(7), 3–7.

Cooke, M., & Matarasso, B. (2005). Promoting reflection in mental health nursing practice: A care illustration using problem-based learning. *International Journal of Mental Health Nursing*, *14*, 243–248.

Dochy, F., Segers, M., Van den Bossche, P., & Gijbels, D. (2003). Effects of problem-based learning: A meta-analysis. *Learning and Instruction*, *13*(5), 533–568.

Farideh Esfandiary, K. (1995). *The effectiveness of problem-based learning in nurse practitioner education*. PhD Dissertation, University of Southern California. Received 19 February, 2007, from Proquest database.

Fenwick, T. J. (2002). Problem based learning, group process and the mid-career professional: Implications for graduate education. *Higher Education Research and Development*, *21*(1), 5–21.

Forneris, S. G., & Peden-McAlpine, C. J. (2006). Contextual learning: A reflective learning intervention for nursing education. *International Journal of Nursing Education Scholarship*, *3*(1), Article 17. Retrieved from http://faculty.ksu.edu.sa/hisham/Documents/Medical%20Education/English/Nursing%20Education/55.pdf

Gillespie, M., & Peterson, B. L. (2009). Helping novice nurses make effective clinical decisions: The situated clinical decision-making framework. *Nurse Education Perspective*, *30*(3), 164–170.

Haggerty, C. (2009). Developing an online course: Reflections on the process. *Whitireia Nursing Journal*, *16*, 8–17.

Holloway, K. (2011). *Development of a specialist nursing framework for New Zealand*. Unpublished Doctoral Thesis, University of Technology Sydney, Sydney.

Jefferies, P. (2007). *Simulation in nursing education: From conceptualization to evaluation*. New York, NY: National League for Nursing.

Kantar, L. (2014). Incorporation of constructivist assumptions into problem-based instruction: A literature review. *Nurse Education in Practice*, *14*(3), 233–241.

Lateef, F. (2010). Simulation-based learning: Just like the real thing. *Journal of Emergencies, Trauma, and Shock*, *3*(4), 348–352.

McCaughey, C. S., & Traynor, M. K. (2010). The role of simulation in nurse education. *Nurse Education Today*, *30*(8), 827–832.

Mc Millan, D. E., Bell, S., Benson, E. E., Mandzuk, L. L., Matais, D. M., McIvor, M. J. ... Wilkins, K. L. (2007). From anxiety to enthusiasm: Facilitating graduate nursing student's knowledge development in science and theory. *Journal of Nursing Education*, *46*(2), 88–91.

Offricht, K. (2012). The power of mindful learning. *Reflections on Problem-Based Learning (PBL)*, *13*, 10–12.

Prince, M., & Felder, R. (2007). The many faces of inductive teaching and learning. *Journal of College Science Teaching*, *36*(5), 14–20.

Profetto-McGrath, J., & Worrell, J. (2010). Nursing research in Canada. In B. Kozier, G. Erb, A. Berman, S. Snyder, D. Bouchal, S. Hirst ... M. Buck (Eds.), *Fundamentals of Canadian nursing concepts, process and practice* (2nd ed.). Toronto: Pearson Education Canada.

Rudolph, J., Simon, R., & Raemer, D. (2007). Which reality matters? Questions on the path to high engagement in health care simulation. *Society for Simulation in Healthcare*, *2*(3), 161–163. doi:10.1097/SIH.0b013e31813d1035

Tanner, C. A. (2006). Thinking like a nurse: A research-based model of clinical judgement in nursing. *Journal of Nursing Education*, *45*(6), 204–211.

Trimmer, W., Laracy, K., & Love-Gray, M. (2009). *Seeing the bigger picture through context-based learning*. Good Practice Publication Grant e-Book. New Zealand: Ako Aotearoa. Retrieved from www.akoaotearoa.ac.nz/gppg-ebook

White, M. J., Amos, E., & Kouzekanani, K. (1999). Problem-based learning: An outcomes study. *Nurse Educator*, *24*(2), 33–36.

EVALUATING EUROPEAN INQUIRY-BASED EDUCATION PROJECTS THROUGH THE LENS OF EVALUATION CAPACITY BUILDING

Fabio Dovigo and Vincenza Rocco[†]

ABSTRACT

Inquiry-Based Science Education (IBSE) is considered as a powerful pedagogical practice for renewing science teaching in and out of school. However, despite the spread of IBSE projects, solid evaluation practices have not yet been adequately developed. Practitioners are often uncomfortable with the evaluation task, usually spending little time on it and using poor tools to evaluate IBSE activities. To investigate this issue, we developed an analysis of the European project INQUIRE within the

[†]Although the chapter has been jointly conceived and discussed by the two authors, Fabio Dovigo is specifically responsible for Sections 1–4, as Vincenza Rocco for Sections 5–8.

Inquiry-Based Learning for Science, Technology, Engineering, and Math (STEM) Programs: A Conceptual and Practical Resource for Educators
Innovations in Higher Education Teaching and Learning, Volume 4, 257–282

ISSN: 2055-3641/doi:10.1108/S2055-364120150000004015

framework of Evaluation Capacity Building theory (ECB). INQUIRE fosters IBSE activities in schools and botanic gardens, promoting training for teachers and educators working with pupils aged between 9 and 14. ECB aims at developing evaluation capabilities in order to conduct rigorous evaluation. Adopting a mixed-methods approach, we systematically analysed evaluations used within courses held in Italy over two years (2011–2013). Study findings suggest that the INQUIRE project has helped practitioners to improve their ability to conduct effective evaluations. However, we also identified two concerns, as IBSE practices tend to focus on cognitive rather than cooperative teaching elements, and on summative rather than formative evaluation methods. The following four recommendations are consequently suggested: promoting dynamic assessment processes, ensuring clearness and flexibility, valuing tacit knowledge, and fostering cooperation.

THE THEME OF THE RESEARCH AND THEORETICAL FRAMEWORK

In recent years, many international studies have underscored a worrying decline in young people's interest in science and mathematics, largely due to the way science is taught in schools (Eurobarometer Report, 2010; PISA, 2009). As one teacher interviewed during the research effectively illustrated, direct instruction is still the most common educational approach to scientific activity, with the teacher narrating textbooks as learners listen passively or answer stock questions. Even when 'experiments' are carried out in a laboratory, the teacher is the central focus of attention, designing and providing procedures step-by-step while students are executing tasks rather than actively posing questions or exploring hypotheses. Conversely, in a constructivist approach to scientific education, the aim is to ease learners into building up their knowledge to learn together, starting with a discussion of preconceptions (Jenkins, 2013). Assuming a facilitator's role, the teacher actively engages students in using flexible strategies to solve real problems, consistent with students' knowledge and abilities. Discussion and documentation are essential to this process (Harlen, 2006). The Rocard Report (2007) strongly recommended the dissemination and integration of innovative Inquiry-Based Science Education (IBSE) methods to foster the constructivist approach to science teaching. IBSE is an active approach that leads students to develop key scientific ideas through learning how to investigate and build knowledge and understanding of the real world (Harlen, 2013). As in professional scientific inquiry, IBSE involves active learning through making observations; posing questions; examining primary literature; planning investigations; using tools to gather, analyse and interpret data; proposing explanations and predictions; and communicating the results (NRC, 1996). As in professional inquiry, the IBSE approach is debateable, contested, iterative, reflexive, and both cooperative and competitive. This is in contrast to an 'ordered vision' of science sometimes presented to students with recipe-style laboratory exercises and a control of variables or fair testing (Levy, Lameras, McKinney, & Ford, 2011). In IBSE, the learning process is a cycle using the BSCS '5E' model: Engage, Explore, Explain, Elaborate and Evaluate (Bybee et al., 2006). Each phase has a specific function: to promote curiosity and elicit prior knowledge, to discover data and formulate hypotheses, to demonstrate conceptual understanding and process skills, to develop a deeper and broader understanding and adequate skills and to assess understanding and abilities achieved by the students.

IBSE has emerged as an effective pedagogical practice for revamping science teaching. Several studies have supported its efficacy in improving both teachers' motivation and students' interest and achievement (Duschl & Grandy, 2008; Minner, Levy, & Century, 2010). To spread IBSE, the 7th European Programme, 'Science in Society', has funded innovative projects such as Pathway, Profiles, Pri-Sci-Net and INQUIRE (Bromley et al., 2013).

Compared to the large number of projects based on the IBSE method, the evaluation of inquiry-based activities still remains rather underdeveloped. While many assessment tools for evaluating the quality of IBSE activities and students' skills are currently employed, since IBSE has proved to be a complex process, there is no 'right way' to assess it (Dillon, 2012). As a consequence, despite the fact that 'Science in Society' projects include programme monitoring and formal evaluation to determine whether the intended findings are being achieved, there is a lack of effective programme evaluation as well as a lack of confidence among professionals in their ability to use evaluation in their programs (Coyle, 2005). Moreover, an additional challenge is the inability of organizations to provide resources to support educators in improving their skills in and knowledge of programme evaluations (NEEAC, 1996). To investigate this issue, we developed an analysis of the European project INQUIRE (Inquiry-Based Teacher Training for a Sustainable Future) within the framework of Evaluation Capacity Building theory (ECB).

INQUIRE is a three-year project co-ordinated by Innsbruck University Botanical Garden, involving 17 partners from 11 European countries (Austria, Belgium, Bulgaria, France, Germany, Italy, Norway, Portugal, the United Kingdom, Russia, and Spain). INQUIRE seeks to reinvigorate IBSE in schools, museums and botanic gardens, fostering the professional development of teachers and educators working with pupils aged between 9 and 14 on the topics of biodiversity loss and climate change. Fourteen botanic gardens have developed two 60-hour training courses on IBSE for practitioners and a 'train the trainers' course.

ECB is the intentional work of creating and sustaining organizational processes that make quality evaluation and its uses routine. It involves the supply of technical skills, tools and resources to produce effective and useful evaluations that become sustainable over time (Fleming & Easton, 2010). In order to develop a sustainable ECB, it is essential to embed the evaluation into daily work practices and policies, which enable the promotion of cultural change in schools. This target has been specifically stressed by the teachers and educators who participated in the INQUIRE project.

Building the evaluation capacity of individuals and groups means knowing and discussing their motivation for engaging in ECB, as well as the assumptions and values supporting evaluation, the goals of the assessment practices, and how they contribute to effective decision-making and add value to the school's organization.

EBC fosters the active cooperation of the stakeholders involved through hands-on and learning-by-doing activities planned to design, implement and manage evaluation projects in accountable ways (Preskill & Boyle, 2008). We choose to examine the monitoring process and the assessment tools implemented by the INQUIRE project, through the lens of ECB, to evaluate the ability to provide staff with skills and sufficient resources to conduct rigorous and lasting evaluations.

METHODOLOGY

The investigation involved the analysis of documentation (forms, questionnaires, reports, manuals, and lesson plans) used during the INQUIRE project, research about the stakeholders' opinions and practices (interviews, focus groups), and direct participation in courses, workshops, and meetings held during the project.

In particular, we chose to examine two IBSE training courses for practitioners (teachers and botanic garden educators), which took place in Italy between 2011 and 2013. The first course involved 14 educators and 13 teachers (5 from primary schools, 8 from secondary schools). The second course involved 26 educators and 19 teachers (7 from primary schools, 12 from secondary schools). In addition, we analysed the main evaluation tools used during training courses carried out by the other European partners involved in the INQUIRE project.

To examine the evaluation practices used during the courses, we adopted a mixed-methods approach in order to investigate the following questions:

- Which criteria are most suitable for effective assessment of IBSE activities?
- To what extent and how did practitioners develop evaluation skills through the training courses and implement them subsequently in the classroom and/or in learning outside the classroom practices?

We chose to employ a mixed-methods approach, as the triangulation of quantitative and qualitative tools provides for a comprehensive understanding of the participants' experience and viewpoints. Participants were: 32 teachers from primary and secondary schools; 40 educators from the botanic gardens, museums and parks; 800 students involved in IBSE activities who completed the self-evaluation form; 10 European and Italian teacher trainers involved in the INQUIRE project.

Quantitative data were collected through pre- and post-training course questionnaires, structured interviews and the evaluation form. Qualitative data were collected using semi-structured interviews, participant observation, research journals, world café,[1] concept cartoon,[2] final reports, self-evaluation forms, and portfolios of evidence.

To analyse data, we used SPSS© to carry out statistical analysis of quantitative variables, and interpretative phenomenological analysis (IPA) to detect and examine qualitative variables arising from the participants' experience (Smith, Flowers, & Larkin, 2009). IPA allowed us to explore how teachers and educators ascribe meaning to the assessment practices both in their educational activities and during the training courses. ATLAS-TI© was used to manage the coding and linking of different variables, and to ensure a constant comparison of data in order to improve the reflective process of analysis.

INQUIRE PROJECT: THE EUROPEAN FRAMEWORK FOR EVALUATION

Quality management (QM) within the INQUIRE project has been organized to ensure the effective use of available resources and reliable management of the goals pursued. The main tasks of QM were:

- to build the evaluation plan as well as to provide formative and summative assessment of the courses and the professional development of participants;
- to support the training of the partners regarding the implementation of the evaluation process;
- to assist the partners through meetings with experts held at the botanic gardens involved in the project;
- to help share reflective practices with partners at the European level, and foster the same practices among course trainees at national and local levels.

To achieve these goals, QM prepared four questionnaires (two for the teachers, two for the educators) to be administered at the beginning and end of each course. Additionally, QM devised a set of guidelines to formulate the portfolio of evidence as a tool for analysing, documenting, and evaluating the project process and outcomes. Moreover, QM prepared a chapter focused on assessment methods within the practitioners' handbook for teachers and educators, clarified the difference between formative and summative evaluation both in theoretical and practical terms, and offered a range of tools and techniques to evaluate students' learning, including observation, research journals, and the World Café.

Special attention was paid to reviewing the evaluative practices adopted during the first training course, asking the participants to illustrate what assessment tools proved to be the most effective, for what reasons, and how to implement them in future course activities. The final picture showed that each country had employed a diversified and particular range of tools in order to adapt them to the cultural and local features of the course. As a consequence, QM was careful not to impose a strict standard framework, enabling each partner to make use of the set of tools they deemed most appropriate to the context in which they operated.

In short, participation in the project meetings stimulated the members to assess, evaluate and improve the training courses and, more generally, the activities promoted within the INQUIRE framework, on the one hand

Table 1. Overview of INQUIRE Evaluation Instruments Used at the European Level.

Instruments		Partners
Asking questions	Questionnaire	NBGB, BelgiumUBG, BulgariaUAH-CSIC, SpainBGRHB, GermanyBORD, France
	Interview	MSU, RussiaBGRHB, GermanyUni-HB, GermanyLFU, Austria
Observing	Observation (paper and pencil)	UBG, BulgariaKEW, United KingdomBGRHB, Germany
	Observation (grids)	UL, Portugal
	Video recording	SBZH, Germany
	Social Network Analysis	LFU, Austria
	Energy Graphics	UBG, Bulgaria
Schematizing	Conceptual Maps	LFU, Austria
	Mind Maps	SBZH, Germany
Discussing	World Café	MUSE, Italy
	Open Group Debate and Discussion	KEW, United Kingdom BORD, France
	Personal Meetings	MSU, Russia NHM, Norway
Reflecting	Portfolio of Evidence	MSU, Russia UL, Portugal FCTUC, Portugal BORD, France MUSE, Italy
	Final Report	UAH-CSIC, Spain

assuming a habit of constant reflection about the educational actions undertaken, on the other openly sharing outcomes, successes and shortcomings to favour the continuous professional improvement of every partner.

In light of the collected data, tools were eventually grouped into five sets: asking questions, observing, building mind and conceptual maps, discussing and reflecting. The various sets represent the different research paths possible to take through the evaluation process (Table 1).

EVALUATING ITALIAN TRAINING ON IBSE METHODS USING THE ECB LENS

The Italian project staff opted for several of the tools proposed by the QM to evaluate the IBSE training courses. Quantitative and qualitative methods were used to collect the participants' opinions before, during and after the course to monitor the field activities carried out at the botanic gardens and to evaluate the learning outcomes of the students participating in the project.

The selected tools were: questionnaires (pre- and post-); semi-structured and Delphi interviews,[3] research journals (during training); and World Café (at the end of the course).

The Questionnaires

Within the INQUIRE project, questionnaires have been developed to collect information about trainees' opinions, expectations and practices and to evaluate the impact of the project on the participants' professional growth. Through the online platform Questback©, the organizers administered a total of eight anonymous questionnaires, addressed to the teachers and educators at the beginning and the end of the pilot and mainstream course, respectively. As open questions dominated in the first questionnaire, only a small number of questionnaires were returned after first course, a completion proved generally too laborious for participants. Consequently, the organizers decided to structure the second course questionnaire mainly around closed questions to raise the completion rate by reducing the time required to complete the form.

Overall, at the end of the project 96 questionnaires were collected (Tables 2 and 3).

Table 2. Respondents to the Questionnaires (First IBSE Course).

Course 'IBSE: like a scientist!' (Italy)			
	Teachers	Educators	Total
Pre-training Questionnaire	12 (92%)	8 (54%)	20 (74%)
Post-training Questionnaire	8 (62%)	2 (14%)	10 (37%)

Table 3. Respondents to the Questionnaires (Second IBSE Course).

Course 'IBSE: like a scientist!' (Italy)			
	Teachers	Educators	Total
Pre-training Questionnaire	18 (95%)	22 (85%)	40 (88%)
Post-training Questionnaire	15 (79%)	11 (42%)	26 (58%)

Analysis of the answers related to the topic of evaluation demonstrated that teachers and educators shared a strong interest in improving their assessment skills (Fig. 1). Trainees indicated ability to evaluate the efficacy of educational interventions as a crucial competence for qualified teachers and educators (Fig. 2).

Compared to those filled out at the beginning of the course, post-course questionnaires showed that participants thought that the project greatly helped them to improve their skills to manage educational evaluation. Even though they felt that a large gap remained to be filled before they could consider such an ability permanently acquired (Figs. 3 and 4).

With reference to the implementation of the five evaluation phases envisaged by the IBSE approach, the Evaluate step proved to be less developed compared to the others (Figs. 5 and 6).

Teachers and educators pinpointed that the five-phases model was not clear in terms of differences between one step and another. Moreover, time requirements for each phases were seen as not fully specified. They suggested that the model needed to be clarified, especially through examples of what represents appropriate implementation and feedback on trainees' attempts to put it into practice.

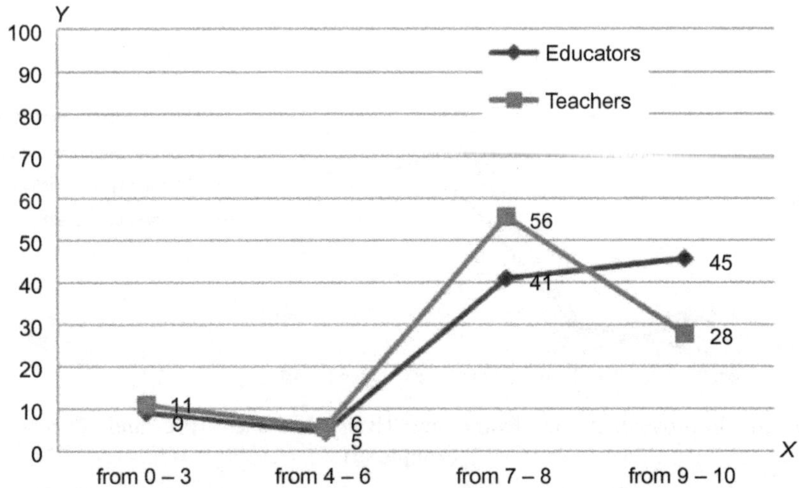

Fig. 1. Improvements in Evaluating IBSE Activities.

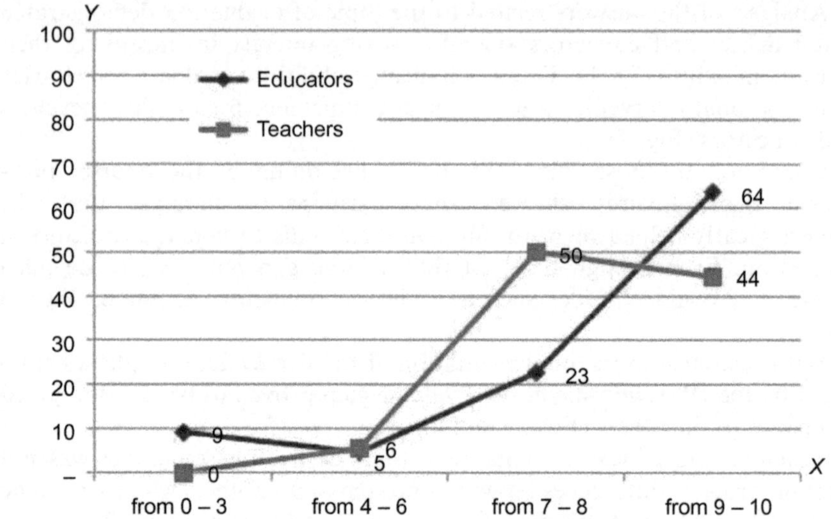

Fig. 2. Improvements in Evaluating the Efficacy of IBSE Lessons.

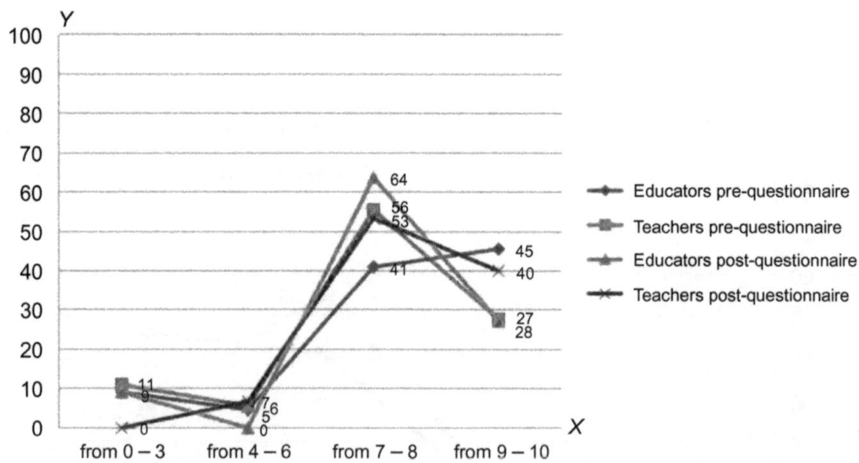

Fig. 3. Improvements in Evaluating IBSE Activities (Pre and Post-course Compared).

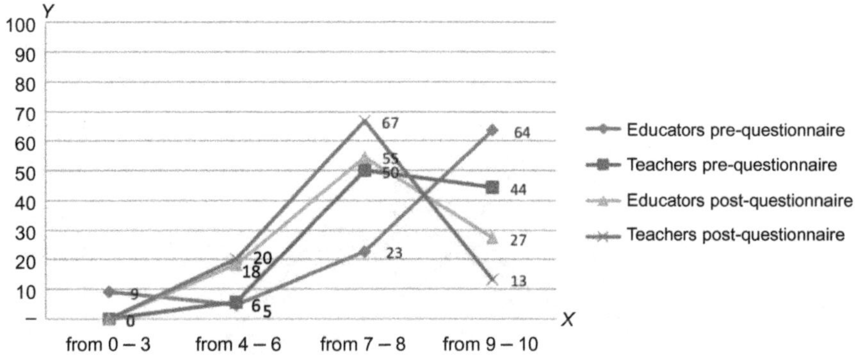

Fig. 4. Improvements in Evaluating the Efficacy of IBSE Lessons (Pre and Post-course Compared).

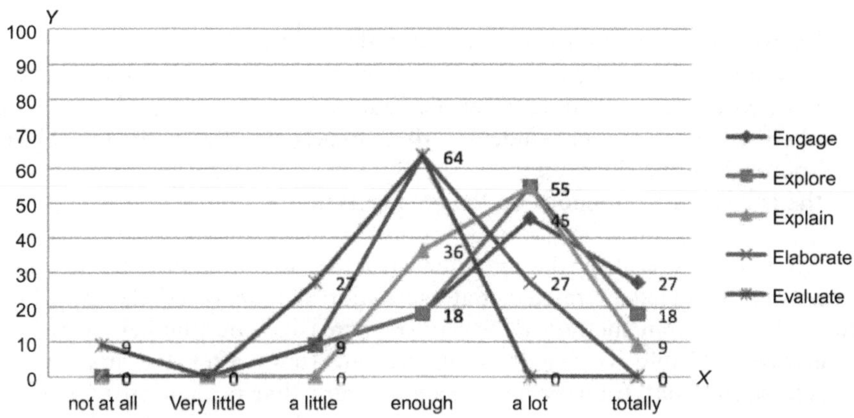

Fig. 5. The 5 E Model Implementation during the IBSE Activities (Educators).

Semi-structured and Delphi Interviews

Semi-structured interview is an approach based on a fairly open framework which allows for focused, conversational, two-way communication. It was used to collect the participants' opinions during the two training courses, so as to get immediate and ongoing feedback from the participants. Questions were asked to obtain information about:

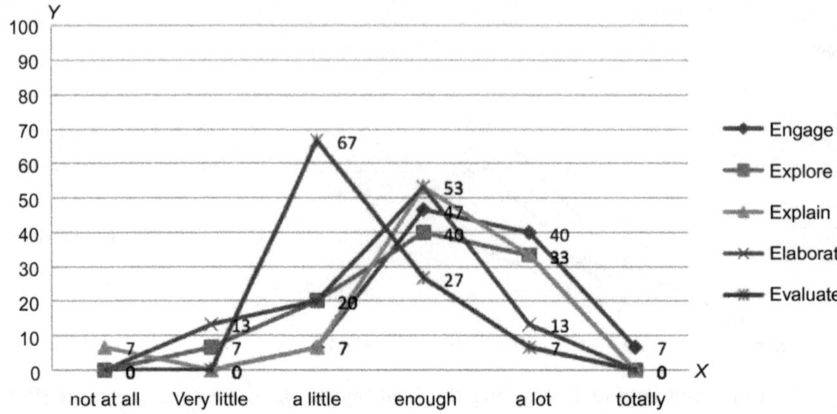

Fig. 6. The 5 E Model Implementation during the IBSE Activities (Teachers).

– knowledge and skills acquired through the training;
– how much the participants modified their teaching styles and interactions with colleagues because of the course;
– observed changes in students after experimenting with the IBSE approach in the classroom;
– the participants' opinions on the course itself.

After the first course, the interview structure was modified. Some questions were changed or refined, but the major shift involved the introduction of the Delphi method as a way of increasing the number of people interviewed (Table 4). Analysis of the interviews after the first year's course shows that trainees had become aware that evaluation needed to be reconsidered, especially as a more distributed practice within the classroom. Moreover, they emphasized that evaluation deserved more time, required more variation in methods, and called for greater attention to the five steps of the IBSE approach (especially the Evaluate phase). Also, the interviews highlighted an emerging sense of anxiety about the final evaluation of the IBSE activities, as trainees thought that it should be different from traditional evaluation. Two participants commented as follows:

> First of all I learnt to pay attention to the different evaluation steps. Before I didn't think a lot about it during my classes. The course gave me the opportunity to reflect and carefully articulate the assessment phases.

> What I am really scared of is the final evaluation ... Afterward the IBSE course it should not be the traditional evaluation we usually did anymore, we need to involve the students more directly in reflecting on the work they did.

Trainees also stated that involving the students in the evaluation process was quite demanding compared to what they normally do in class, as it requires a huge amount of time and the ability to master complex tools.

The Research Journal

The research journal was employed throughout the course as a means of getting feedback from the participants in terms of reflections and evaluation. At the end of each module, trainees were invited to answer two to four open questions about the topic. Even though the number of research journals collected at the end of the first course was low (19%), the journal was proposed again during the second course, emphasizing the value and meaning of this tool. As a result, the number of research journals kept by teachers and educators at the end of the second course increased markedly (44%, Table 4).

A number of critical points had already emerged from the first research journals that criticized evaluation practices proposed during the course, and stressed the lack of clarity about the distinction between skills and content evaluation. Consequently, during the second course, the organizers decided to spend more time and resources delving into the educational reasons and goals of evaluation, and the use of the assessment tools. The analysis of research journals at the end of the second course highlighted a shift in trainees' perceptions of evaluation. Evaluation was seen as a professional tool to reflect on educational action, and formative evaluation had become central both as a way to avoid the linear assessment of children's skills development and a useful tool enabling teachers and educators to question their pedagogical decisions.

One teacher wrote:

> Evaluation is a process that helps you to gain feedback on your work. In fact through evaluation I can understand how much content and knowledge has been transmitted to the children and also how effective my teaching methods have been. I believe that the way in which a lesson is taught should also be evaluated, and its dynamics and possible variations studied.

Two additional educators noted:

> Evaluation is a way of understanding whether we were effective in doing the planned activity and how.

> Evaluation allows you to understand what remains after putting time and effort into a certain educational path.

Table 4. Respondents to the Evaluation Instruments (Pilot Course and Mainstream Course).

	Pilot Course	Mainstream Course	
	Interview	Interview	Delphi Interview
Teachers	5 (38%)	7 (37%)	7 (37%)
Educators	2 (24%)	3 (12%)	5 (19%)
Total	7 (26%)	10 (22%)	12 (27%)
	Research journal	*Research journal*	
Teachers	5 (38%)	11 (58%)	
Educators	0	9 (35%)	
Total	5 (19%)	20 (44%)	
	Evaluation form		
Teachers	7 (54%)		
Educators	1 (7%)		
Total	8 (30%)		
	Final report		
Teachers	9 (69%)		
Educators	3 (21%)		
Total	12 (44%)		
		Post-experimentation form	
Teachers		11 (58%)	
Educators		7 (27%)	
Total		12 (44%)	

At the end of the course, evaluation was seen not only as a way of checking the knowledge acquired, but also as a kind of inquiry into the students' ability to elaborate on their knowledge and to transfer it to other educational domains.

The World Café

As we said, the World Café is a public meeting that facilitates cooperation between people, stimulating active contribute ideas, to link together their findings, to interconnect individual and collective dimensions and to build knowledge through creative thinking and through sharing problems that arise from open questions (Brown, Isaacs, The World Café Community, 2005). This technique perfectly matches the IBSE approach in emphasizing the role of active cooperation and idea-sharing among the participants involved in the project.

The World Café was employed during the course both to develop new activities through creative workshops, evaluate the evolution of the training by asking for the participants' opinions about the strong and weak points they observed, and elicit their suggestions about possible improvements to the training experience.

Through the Café, trainees pointed out that evaluation generally suffers from a negative reputation in school and teachers often feel unprepared as they deal with this specific task. Moreover, participants complained that evaluation did not receive adequate time and attention compared to other activities scheduled for the course. Consequently, they urged the organizers to delve more deeply into this topic during the second part of the program. The latter approved this request, organizing a training section devoted to the implementation of evaluation practices and tools. Moreover, they modified the schedule of fieldwork that had been planned for later in order to leave teachers and educators more time to complete the evaluation tasks.

THE ANALYSIS OF IBSE EVALUATION TOOLS

IBSE activities have been analysed using a number of tools, including evaluation forms and final reports (for the pilot course), post-test and observation forms (for the mainstream course) and the self-evaluation form. Evaluation involved different IBSE actors, namely, teachers, educators, students and external observers.

The Evaluation Form

The evaluation form was conceived with the objective of helping trainees to develop self-evaluation and analyse students' behaviour during the IBSE activities. External observers used the same form to analyse teachers, educators and students interacting during classroom and fieldwork activities, to point out positive and negative events, and to provide feedback for improvements.

Only a small portion (27%) of the participants used this form, namely, seven teachers, one educator and three external observers (Table 4). Even though few forms were actually collected, it is interesting to compare them to the related lesson plans, which were the operational programs conceived to organize the IBSE activities step-by-step. The comparison highlights that most of the teachers thought that evaluation was a useful means of observing and assessing several educational elements such as:

- relational abilities;
- teamwork effectiveness;
- planned sequence of activities;
- the ability to explain logically;
- the specific knowledge acquired about the lesson topics;
- the final individual and/or group outcomes of the activity;
- the ability to link everyday and school experiences.

Therefore, beyond assessing scientific knowledge, practitioners assessed students' inquiry skills as raising questions, observing, designing investigations, collecting data, linking evidence to explanations, finding appropriate solutions, listening to each other, getting along socially, and working cooperatively. Analysis of evaluation forms shows general agreement about the value of helping students to develop self-evaluation. Conversely, not all trainees agreed about the efficacy of peer evaluation. A shared concern was the teachers' inclination to focus attention on the individual student, a tendency reinforced by the use of observation grids. Accordingly, the analysis suggests it would be beneficial to actively involve students both in self-evaluation and in the assessment of other students, provided that teachers and educators ensure that critiques are fair and help to promote the reflective process. In fact, the ability to help students to acquire a cooperative approach in reviewing each other contributions to the discussion cannot be taken for granted. Teachers should help students to readily recognize and value what they see as a positive and supportive contribution.

The Final Report

The final report was conceived as a set of categories implemented to categorize the course outcomes, that is, as a metacognitive tool for facilitating self-evaluation and to help teachers and educators to reflect on the experimentation carried out throughout the IBSE project. We focused the analysis on seven key points: cooperative learning, teacher as facilitator, students active participation, knowledge and skills acquired by students, the 5E, assessment process and teaching approaches.

As shown in Table 4, 12 final reports were written representing 44% of trainees. Participants commented positively on the reflective stance the course helped them to assume on evaluation:

> The course enabled me to reflect on the steps we need to follow during classroom activities, and especially to pay attention to the Engagement and Evaluation phases. (Teacher, Primary school)

At the same time, teachers and educators seemed to acknowledge evaluation as a difficult and thorny (sometimes problematic) task:

> It is not easy to say whether and what kind of knowledge we have acquired ... For sure, we developed skills in team-working, sharing ideas, organizing work plans and making hypotheses. Maybe some gained more complex skills than others. But for sure student motivation has improved. Therefore an important goal has been reached!. (Teacher, Lower secondary school)

Other important remarks emerging from the final reports are the following:

- the importance of openly praising students when they are showing what they have discovered and learned through the IBSE experience;
- the significance of seeing evaluation as an ongoing and recursive activity, in which observation, reflection and assessment are part of a holistic process deeply embedded into the students' educational path;
- the need to discriminate between evaluation of content (to know something) and evaluation of competences (to be able to do something);
- the value of becoming able to use a wide range of evaluation tools such as conducting observations, posing questions, formulating explanations, making predictions, planning and conducting scientific investigations, interpreting data, communicating observations and results, and working in groups.

The Post-experimentation Form

In the second IBSE course, a new form called 'post-experimentation form' replaced the previous evaluation form and the final report. The new form invited trainees to self-evaluate the IBSE activity they designed, observing students in the classroom and during fieldwork, and reflecting on the competencies developed and the evaluative tools adopted.

The post-experimentation form was filled out by 40% of trainees. Participants thought that they correctly put the IBSE approach into action and showed genuine enthusiasm for the INQUIRE project (Figs. 7 and 8). Educators expressed very positive opinions about the activity they completed, whereas teachers remained more cautious about it (Fig. 9).

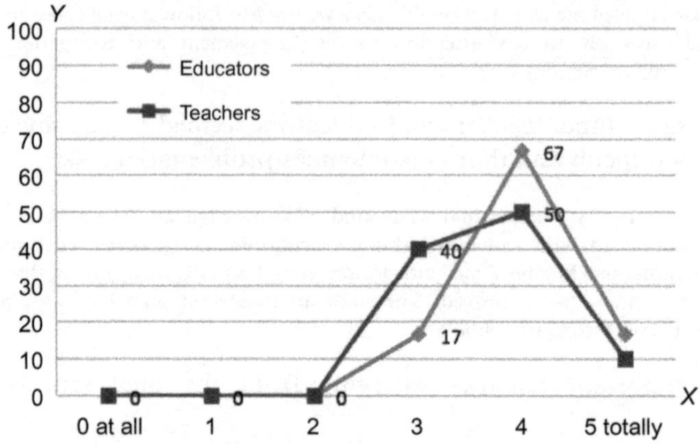

Fig. 7. Implementation of the IBSE Activities.

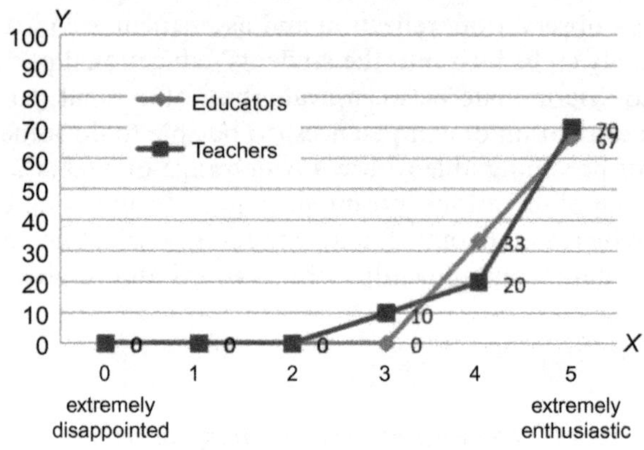

Fig. 8. Enthusiasm at the End of IBSE Activity.

According to the participants, some elements needed to be improved: the timing of the activity, a clearer definition of the activity's goals and the application of the 5E model to the activity's content and students' potential. Additionally, in the forms collected, teachers emphasized the following as essential priorities that should be more carefully evaluated: linguistic, communication, relational and planning skills; the ability to produce questions, observations and hypotheses; the role of team-work; the processing of new concepts and the transfer of acquired knowledge to new contexts.

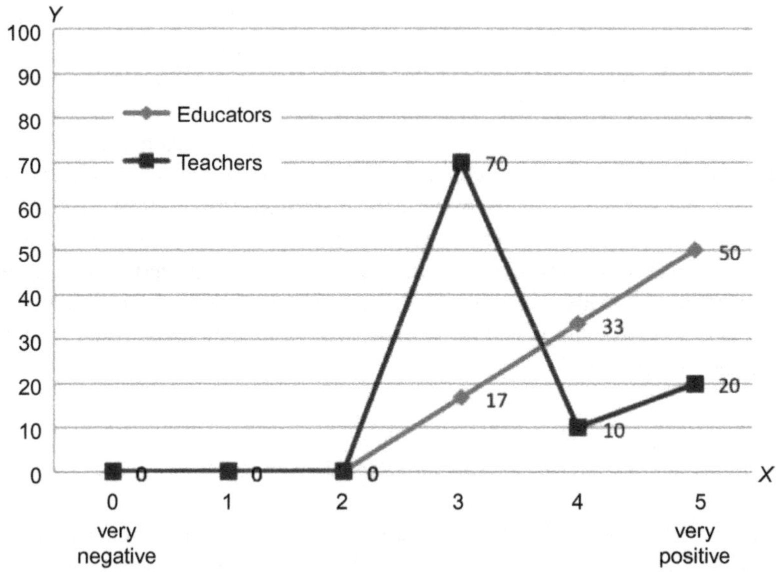

Fig. 9. Evaluation of IBSE Activity.

The educators stressed the need to build more competencies in conducting research as well as fostering the participation and real curiosity of students. Moreover, as most of the tools adopted to ensure the assessment of the IBSE activities were unfamiliar to trainees, they suggested that the organizers should allocate more time to allowing participants become acquainted with them in the future.

All participants agreed that, even though the course had triggered their interest in developing evaluative skills, they needed even more training to delve into the difference between summative and competence evaluation, as well as to become able to promote a wider culture of evaluation in schools.

Observation

During the second course, the organizers suggested observation as a means for supporting participants through the analysis of the five phases envisaged by the IBSE approach. Observation was focused on the behaviours and interactions among students, teachers and educators in order to highlight critical elements and evaluate the way the approach was actually implemented. Even though, for organizational reasons, only few observations were fully achieved, the data highlighted two interesting points. On the one hand, observers noted that trainees paid only marginal attention to evaluation during activities, often postponing it to the end of the educational process. Evaluation was very often identified with completing a final task, as well as confined to students' oral comments prompted by teachers' open questions posed at the end of the project. On the other hand, observers criticized the automatic association which often occurred between evaluation and teaching. Indeed, they suggest that an effective evaluation process requires more active involvement on behalf of all people involved in the project, and especially self- and peer evaluation of students.

Self-evaluation Form

Students evaluated themselves by answering questions concerning what they learned through the IBSE activity, including positive discoveries, barriers encountered and the experience of working in a group.

Even though only a small portion of trainees (7% in the pilot course, 29% in the mainstream course) filled out the form, the respondents found it useful in providing relevant feedback about scientific knowledge, attitudes and social skills acquired by students. All in all, through the form, trainees were able to gather more evidence to improve IBSE activities, and students had a stronger say in the evaluation process.

Tools to Assess the Students' Learning Outcomes

Concept maps, portfolios of evidence and concept cartoons were introduced to practitioners through examples and practical activities during the training course to help trainees both observe and evaluate student learning outcomes.

- *Concept map:* Although concept maps are a well-known tool for learning and evaluation, only one educator and one teacher implemented them during the course as a way to analyse students' previous knowledge and assess learning. Practitioners generally found it difficult to integrate maps in their daily educational practices, as they considered them as time-consuming activities.

- *Portfolio of evidence*: Neither during the first nor the second course the portfolio was awkward to follow, as trainees thought that it was too demanding, requiring considerable time to document, analyse and reflect on ongoing practices. Only one teacher wrote an accurate portfolio, even though it was focused more on describing activities than developing critical reflection. Again, teachers found this instrument too laborious to fit with the complex flow of everyday activities. Conversely, the staff was able to effectively implement the portfolio approach through focusing on more limited and discrete events as those represented by trainees' learning classes.
- *Concept cartoons*: Practitioners really liked concept cartoons as a quick, simple and effective means of evaluating students' learning. Three trainees designed and implemented cartoons focused on floatage, seed dispersal and sustainable diet to, respectively, provoke discussions, stimulate thinking and evaluate the development of social skills and scientific knowledge during the IBSE project.

DISCUSSION

Our research shows that the INQUIRE project contributed to building evaluation capacity in professionals, providing convenient resources and fostering technical skills that helped practitioners to learn about evaluation.

The main outcomes concern cognitive, behavioural and affective aspects, including:

- an increased understanding of evaluation as a multidimensional approach, including the awareness of the difference between summative and formative assessment, and the importance of adopting a reflective stance during evaluation;
- an enhanced ability to use a plurality of assessment tools (such as World Café, concept cartoons and portfolios) to collect and combine different types of data about environmental education;
- decreased evaluation anxiety and increased confidence in the usefulness of evaluation practices built on continuous networking among professionals.

Both teachers and educators think that evaluating students is actually a crucial but difficult task, as the relationship between experience and learning often revealed to be more indirect and elusive than expected. They appreciate the structure of IBSE as paying special attention to the evaluative process. However, unlike in the first training course, in the second one, they were able to spend more time clarifying the meaning and purposes of evaluation and putting systematic assessment into practice through the use of concept cartoons, forums, diaries, questionnaires, concept maps, portfolios of evidence, interviews and observations.

Concept cartoons were especially appreciated as creative combinations of words and drawings enabling teachers and pupils to explain their perspectives about IBSE activities. As a teacher stated during the follow-up interview:

> I really like concept cartoons. I introduced them in my class as a basis for debating environmental subjects.

The IBSE forum also proved to be a valid source of support for developing an evaluation culture, especially because it allows teachers to share opinions, experiences and questions about how to implement assessment effectively.

Teachers and educators agree on the significance of listening to different opinions and sharing experiences in order to improve knowledge and develop new creative ideas. In the same follow-up, one educator positively commented about this way of sharing ideas:

> Comparing notes with other teachers and educational staff during the course has been very important. We can always learn from each other's experience, voice doubts, clarify problems and discuss the difficulties we have encountered along the way and formulate new ideas.

Similarly, the research journals helped professionals to actively reflect on their own personal roles in the assessment process:

> We need to evaluate not only the knowledge that the students acquired, but also the inquiry and social skills developed during the IBSE activities, for example working in small groups. Besides, evaluation helps me to understand how successful I have been in managing the IBSE activity.

The questionnaires confirm that combining both external evaluation and self-evaluation is a very effective way of analysing the practitioners' need for continuous training in terms of the IBSE pedagogical approach. The investigation further highlights two additional points.

First, one shortcoming can be pictured as a sort of methodological "funnel" narrowing the focus from cooperation to individual learners. IBSE projects are based on the constructivist view of learning as process of knowledge building based on context and discussion between expert and novices, which emphasizes the cooperative role of learners as a community. However, current evaluation practices used by IBSE are only partially consistent with this participative model, because, in reality, a cognitive model, oriented towards a more individual learner perspective, often prevails. This individual perspective is particularly emphasised by the use of observation grids strictly focused on the behaviour of one student at a time, and multiple choice tests or true–false tests that assess the information the students have acquired. Teachers generally comply with this attitude; it is a shortcut for coping with the pressure of assigning grades and offering quick evidence of learning improvements. As a teacher noted:

> I disagree with the current spread of bureaucratic evaluation based on the quantity of information a child is supposed to swallow. Assessing skills takes time.

Cooperative practices such as collecting data in a group, planning and conducting investigations, interpreting evidence, drawing conclusions and discussing results with peers and adults have to be taken into account, as they are at the centre of the IBSE approach. An ECB perspective can help to provide a more balanced evaluation style that can further enhance the benefits of the IBSE participative model.

The second critical point can be imagined as a type of 'black box'. Although one purpose of INQUIRE has been to introduce a positive view of evaluation practices, professionals still find it hard to see evaluation as a supportive process for learning activities. Evaluation procedures are often fragmentary or confined to limited periods of time. They are mainly focused on pre- and post-tests more than on the real dynamics of educational interplay. This in-between process is seen as a black box. As a consequence, teachers and educators usually see assessment more as a summative than a formative practice and the learning potential of students is frequently overlooked.

Our research shows that many practitioners are aware that formative evaluation requires careful planning with time and resources. Most of them want to improve their abilities to conduct meaningful evaluation of scientific educational practices. The evaluation activities developed during the INQUIRE project confirm that ECB could be a reliable framework to integrate time constraints with high-quality assessment.

However, a 'suspicious' or even sceptical attitude about the effectiveness of formative assessment sometimes resurfaces, proving that building high levels of confidence among professionals in terms of evaluation is a lengthy process. Fixing the misalignment between the cognitive and constructive paradigm and summative and formative assessment would be an important step in the right direction. Supporting the continuous training of professionals on the basis of the ECB perspective as a way to provide innovative competencies and tools to be implemented in daily evaluation practices.

Our research highlights four points as crucial steps towards overcoming these weaknesses and improving evaluation practices as follows:

- *Dynamic assessment processes*: Systematically gathering information about changes in scientific knowledge and skills over a period of time;
- *Clearness and flexibility*: Sharing evaluation standards with learners and updating them according to course development;
- *Valuing tacit knowledge*: Taking on students' preconceptions as a challenge for building new knowledge;
- *Cooperation*: Fostering debate based on different viewpoints to develop students' metacognition about acquired knowledge and skills.

LIMITATIONS OF THE STUDY

This study has some limitations that must be pointed out. First of all, considering the wide range of evaluation tools employed during the course, the amount of information and answers provided by trainees was lower than expected. This can be interpreted as a manifestation of participants' uneasiness with regard to the topic of evaluation and its role in developing effective educational strategies. Second, we initially planned to extend the original sample of the subjects – based on Italian course trainees – to other IBSE courses held in the rest of Europe; however, this was only partially implemented due to time constraints.

FUTURE RESEARCH

Our research suggests that adopting the ECB approach in IBSE projects would be beneficial in terms of gaining a broader understanding of the role of evaluation in science education, and the way systematic evaluation actually impacts educational practices. Moreover, we suggest future research including a longitudinal study should focus on the long-term effects of ECB on the way practitioners develop their ability to employ evaluation during IBSE activities.

NOTES

1. The World Café is a conversational process for prompting conversations linked and built on each other as people move between groups, cross-pollinate ideas, and discover new insights into the questions or issues that are most important in their life, work or community.

2. Concept cartoons are cartoon-style drawings that put forward a range of viewpoints about a science concept or everyday event. This strategy takes account of constructivist views of learning – taking students' ideas into account when planning teaching. By presenting a number of possible alternatives, 'cognitive conflict' generates conditions for learning readiness. It also draws on research into common areas of misunderstanding in science.

3. The Delphi method is a forecasting method based on the results of surveys sent to a panel of experts. Two or more rounds of surveys are sent out, and the anonymous responses are aggregated and shared with the group after each round. The Delphi method seeks to aggregate opinions from a diverse set of experts, without organizing a physical meeting.

REFERENCES

Bromley, G., Regan, E., Kapelari, S., Dillon, J., Vergou, A., Willison, J., & Bonomi, C. (2013). The INQUIRE course manual. BGCI, London.

Brown, L., Isaacs, D., & The World Café Community. (2005). *The world café. Shaping our futures through conversations that matter.* San Francisco, CA: Berrett-Koehler Publishers.

Bybee, R. W., Taylor, J., Gardner, A., Van Scotter, P., Powell, J. C., Westbrook, A., & Landes, N. (2006). *The BSCS 5E instructional model: Origins and effectiveness.* Colorado Springs, CO: BSCS.

Coyle, K. (2005). *Environmental literacy in America: What ten years of NEETF/Roper research and related studies say about environmental literacy in the U.S.* Washington, DC: The National Environmental Education & Training Foundation.

Dillon, J. (2012). Panacea or passing fad, how good is IBSE? *BGCI Review, Roots*, *9*(2), 4–8.

Duschl, R. A., & Grandy, R. E. (2008). Reconsidering the character and role of inquiry science: Framing the debates. In R. A. Duschl & R. E. Grandy (Eds.), *Teaching scientific inquiry*. Rotterdam, Taipei: Sense Publisher.

European Commission. (2010). *Special Eurobarometer report, science and technology*. Retrieved from http://ec.europa.eu/public_opinion/archives/ebs/ebs_340_en.pdf. Accessed on March 31, 2015.

Fleming, L., & Easton, J. (2010). Building environmental educators' evaluation capacity through distance education. *Evaluation and Program Planning*, *33*, 172–177.

Harlen, W. (2006). *Teaching, learning and assessing science* (pp. 5–12). London: Sage Publications.

Harlen, W. (2013). *Assessment & inquiry-based science education: Issues in policy and practice*. Trieste: IAP-SEP.

Jenkins, E. W. (2013). Children and the teaching and learning of science: a historical perspective. *Proceedings of children's perspective on school, teaching and learning. Children's perspective on school, teaching and learning*, October 7–9, 2013, Catholic University of Eichstaett – Ingolstadt, Eichstaett.

Levy, P., Lameras, P., McKinney, P., & Ford, N. (2011). *The features of inquiry learning: Theory, research and practice. Pathway to inquiry-based science teaching, (Deliverable 2.1)*. European Commission: CSA-SA Support Actions, Project Number 266624. Retrieved from http://www.pathwayuk.org.uk/what-is-ibse.html. Accessed on March 31, 2015.

Minner, D. D., Levy, A. J., & Century, J. (2010). Inquiry-based science instruction – What is it and does it matter? Results from a research synthesis years 1984 to 2002. *Journal of Research in Science Teaching*, *47*(4), 474–496.

National Environmental Education Advisory Council [NEEAC]. (1996). *Report assessing environmental education in the United States and the implementation of the national environmental education act of 1990*. Washington, DC: Environmental Protection Agency; Environmental Education Division.

National Research Council [NRC]. (1996). *National science education standards*. Washington, DC: National Academy Press.

OECD. (2009). PISA (*Programme for International Student Assessment*). Report: What students know and can do. Retrieved from http://www.oecd.org/pisa/pisaproducts/48852548.pdf. Accessed on March 31, 2015.

Preskill, H., & Boyle, S. (2008). A multidisciplinary model of evaluation capacity building. *American Journal of Evaluation*, *29*(4), 443–459.

Rocard, M., Csermely, P., Jorde, D., Lenzen, D., Walberg-Henriksson, H., & Hemmo, V. (2007). *Science education now: A renewed pedagogy for the future of Europe*. Brussels: Directorate General for Research, Science, Economy and Society.

Smith, J. A., Flowers, P., & Larkin, M. (2009). *Interpretative phenomenological analysis: theory, method, research*. London: Sage.

INQUIRY-BASED LEARNING IN MATHEMATICS

Mindy Capaldi

ABSTRACT

Most college students are required to take at least one mathematics course. Many of these students view mathematics as a dry and tedious subject, where the main task is to "plug and chug" using formulas. In contrast, mathematicians see mathematics as a creative process in which real joy comes from grappling with difficult problems and (hopefully) solving them. In this way, mathematics is like a fun puzzle. The challenge is to get students to view mathematics the same way that their teachers do. Inquiry-based learning (IBL) can help solve this problem. The Academy of Inquiry-Based Learning describes IBL as a pedagogical method that encourages students to conjecture, discover, solve, explore, collaborate, and communicate (What is IBL? (n.d.). Retrieved from http://www.inquirybasedlearning.org/?page=What_is_IBL). With IBL, teachers do not lay out all of the formulas and theorems as previous knowledge. Nor do they provide perfect, easily worked through examples and proofs for every new topic. Instead, IBL courses demonstrate the creative process that is mathematics. IBL makes class more enjoyable for both teachers and students, and can bring students closer to the real experiences of mathematicians.

Inquiry-Based Learning for Science, Technology, Engineering, and Math (STEM) Programs: A Conceptual and Practical Resource for Educators
Innovations in Higher Education Teaching and Learning, Volume 4, 283–299

ISSN: 2055-3641/doi:10.1108/S2055-364120150000004016

A HISTORY OF IBL IN MATHEMATICS

The Early Days

Inquiry-based learning in collegiate mathematics derives largely from the "Moore Method" (Frantz, 1983). R.L. Moore was a successful mathematics professor who disapproved of lecture and favored a teaching method in which students proved the course's theorems on their own. As a student, near the end of the 19th century, Moore was likely frustrating to his mathematics teachers. It was clear that Moore was talented in the subject, but he did not fit into the traditional mold of how people taught and learned mathematics. Often, he did not do what he was told and may not have even paid attention to the telling. Even though he was greatly curious, and went through textbooks "like candy" (Frantz, 1983, p. 113), he did not do assigned problems. He disliked lecture, even when the lecturer was a great mathematician. When he did attend lectures, instead of paying attention to what was being said, his mind was working to discover theorems and proofs on his own. After one such experience, he told the lecturer about his internal discoveries, to the frustration of the man who had just explained the same material to the room (Frantz, 1983).

R.L. Moore started as a mathematics student at the University of Texas in 1898 at age 15, and returned to the school as an associate professor in 1920. As a teacher, Moore continued to deviate from the norm. By 1918, his students were remarking upon the lack of books in their classes. They were not given texts, articles, or any previously published material. Moore established a "discover it on your own" rule. If a student violated his rule, he or she was thrown out of the class (Frantz, 1983). To put yourself in the shoes of one of his geometry students, a brief sketch of your typical class day might go as follows:

1. You arrive in class with the work you prepared the night before. You were assigned four theorems to prove, with a given list of axioms to do so. You were able to prove two of them and are nervous about whether you will be asked to prove one of the two unfinished theorems.
2. Professor Moore starts calling on students to do the first proof. Since you are ranked somewhere in the middle out of all of the students, and he starts with the worst student, you have some time before he gets to your name.
3. The first student called on does not have the proof completed. He is skipped. The second student, named John, is ready and goes to present

at the board. You think John has a mistake in the proof, but of course Professor Moore does not point it out. It is up to you to say something.
4. The mistake is fixed, and the class moves on to the next presenter. Once it is time for the third proof, Professor Moore has made it to your name on the list. Luckily, you have this one done! You go to the board and successfully present your work. A student in the audience asks you a clarification question and you *think* you answer it correctly.
5. No student has the last proof finished. It is reassigned for the next class. You are dismissed, without Moore lecturing at all.

In contrast to many contemporary IBL courses, Moore's students were not supposed to collaborate. Moore did not point out flaws or solutions to his students' presentations (Jones, 1977). At times, none of the students could solve a problem. He would not solve it for them. Once when this occurred, Moore not only failed to provide the answer, he instead revealed that it was an unsolved problem in mathematics! Although Moore was strict and made class competitive, he was effective and once they got used to his methods, students typically liked his style. Indeed, former Moore students would complain in their other mathematics classes when the professor did not make them prove theorems on their own (Frantz, 1983). Moore was able to recruit and graduate an impressive number of doctoral students from those who had taken his courses.[1]

Moore's Legacy

Moore managed a very strict classroom. He was considered successful, but there were some drawbacks to his method. Issues such as nonhomogeneity of students, high levels of competitiveness, large class sizes, or needing to teach service courses caused problems for other instructors. Thus, many people now teach using a "Modified Moore Method."[2] One common reform is that the teacher speaks up in class more, asking questions or providing clarification. This can be especially helpful if it is a lower level class with a large variety of student aptitude. Also, the atmosphere of the class may be made more relaxed and less competitive by leaving out such aspects as a ranked list for calling on presenters. A modified Moore course may have multiple students writing proofs or solutions at the board and then explaining them, as opposed to one student at a time.

One popular change to Moore's method was the introduction of a small group structure, which was initiated in large part by Neil Davidson at the University of Wisconsin (Dancis & Davidson, 1970). He wanted to provide average students more of an opportunity to play a part in developing mathematics, even in larger class sizes. Like Moore, Davidson's goal to support an environment similar to that of mathematics research, where students were active and creative. In addition, they learned problem-solving techniques and were trained to communicate the material. Often this involved students discovering or proving theorems, solving problems, or creating examples within their groups, and then presenting their work to the class (Dancis & Davidson, 1970).

Despite these modifications, some aspects of Moore's method remained. Students were given a list of definitions, axioms, theorems, etc. to work from. They were still required to fill in the gaps with proofs or solutions, but they could work through those with other students. Simply looking up a solution was against the rules for both methods (Dancis & Davidson, 1970). In the years since the small group structure was initiated, instructors have continued to revise the method to suit their own challenges and needs. Many of these successful solutions, as well as more details on how to run a typical Modified Moore classroom, will be discussed later in the chapter.

DO WE KNOW IF IBL HELPS MATHEMATICS STUDENTS?

IBL fits into the more general category of active learning in mathematics, which includes other non-lecture-based methods of teaching. Not only has active learning gained popularity, but also research has shown that it is effective. Evidence indicates that active learning, on average, increases student grades and conceptual understanding in science, mathematics, engineering, and technology (STEM) courses (Freeman et al., 2014). Research specific to calculus courses has shown that traditional instruction can lead to low levels of conceptual understanding (Epstein, 2013). Interactive engagement in calculus, when used correctly with quick assessment and feedback to students, results in computational proficiency as least as good as occurs in a lecture class and improves conceptual learning. These results hold even when accounting for class size, instructor experience, textbook, as well as other variables (Epstein, 2013). In 1996, there was enough evidence to convince the National Science Foundation to recommend an inquiry-based approach for STEM classes (National Science Foundation, 1996).

IBL Research Results

When narrowing the focus to study IBL in particular, outcomes are still encouraging, especially given the recent study by Laursen (Laursen, Hassi, Kogan, Hunter, & Weston, 2011). Previous research investigated the effect of IBL on mathematics students within a single course or institution, and found positive results (Kwon, Rasmussen, & Allen, 2005; Springer, Stanne, & Donovan, 1999). The Laursen et al. (2011) study extended the affirmative findings about IBL after evaluating extensive information about students who took a traditional course versus an IBL course at four major universities. The data were broad and varied, including 300 hours of classroom observation, 1,100 surveys, 110 interviews, 220 tests, and 3,200 academic transcripts from over 100 course sections during a period of more than two years. Self-described IBL classes were markedly different from their traditional counterparts. In an average IBL course, over 60% of class time was spent on student-centered activities. In non-IBL courses, 87% of class time was spent on instructors' lecturing. The conclusions of the study provide a strong case in support of IBL (Laursen et al., 2011).

Especially important were Laursen and colleagues' findings concerning previously low-achieving students and women, who both showed statistically significant ($p < 0.05$) gains (2011, pp. 50–54); there were also no negative consequences for other groups. Initially, low-achieving students' grades rose 0.3–0.5 points compared to those of their non-IBL peers, perhaps due to increased problem-solving strategies and better study habits. Those gains lasted through the students' future mathematics courses (Laursen et al., 2011). In non-IBL courses women performed equally to men in terms of grades, but those women reported lower intellectual (conceptual or problem solving) and affective (confidence or interest) gains (Laursen et al., 2011). Other studies have shown similar issues concerning women in math (Good, Rattan, & Dweck, 2012). The gender gap was erased in IBL classes; women were more confident and more interested in taking further mathematics courses, at the same levels as their male peers. Compared to their non-IBL counterparts, female IBL students took, on average, an additional full elective mathematics course (Laursen et al., 2011). In general, IBL students reported higher learning gains and showed altered attitudes and beliefs that more supportive of learning. Inquiry-based students were also more inclined to take additional IBL courses, if available (Kogan & Laursen, 2014; Kwon et al., 2005; Laursen et al., 2011).

The Coverage Issue

A common concern with inquiry-based learning is the coverage issue, defined by Yoshinobu and Jones (2012) as "the set of difficulties that arise in attempting to cover a lengthy list of topics" (p. 303). Mathematics courses, especially those fulfilling a prerequisite, often must cover a certain list of topics that prepare students for their future course(s) or standardized tests. Lecturing allows for scheduling control and can ensure that students are exposed to all of the necessary material. However, going through the material so quickly can have negative effects (Yoshinobu & Jones, 2012).

A year-long study by Schoenfeld (1988) evaluated the lecture-based teaching, and students' scores, of an exemplary high-school geometry teacher. His students performed well on standardized tests. He covered all of the expected topics and excelled in classroom management. He broke problems down into simple steps that made it easy for his students to follow and duplicate. However, when those students were studied further, results showed that their understanding of geometry was only procedural and superficial. They paid attention to the form of an answer, not the content. For example, students could easily understand the structure of a proof, but they did not grasp the logic behind it. The geometry students also developed the belief that all problems should be solved in only a few minutes time, which coincided with their experience in the classroom. This understanding fostered additional views such as the idea that only geniuses are capable of discovering or understanding mathematics (Schoenfeld, 1988). Indeed, these types of beliefs are prevalent among students of many grade levels and a diversity of teachers (Hekimoglu & KittreIl, 2010; Ward et al., 2010; Yoshinobu & Jones, 2012).

A further consequence of focusing on *what* material is covered instead of *how* it is covered is that many students switch away from a STEM major. A sociology study found four main factors that influence students' decisions to leave a Science, Mathematics, or Engineering (SME) major. They are lack of/loss of interest in SME, better education/more interest in a non-SME major, poor teaching by SME faculty, and the curriculum overload or fast pace in SME classes (Seymour & Hewitt, 1997). Instead of having coverage as a top priority, Yoshinobu and Jones (2012) argue that the emphasis should be on "mathematical proficiency," which has five components:

1. Conceptual understanding
2. Procedural fluency

3. Strategic competence
4. Adaptive reasoning (adjusting ideas for other problems)
5. Productive disposition (having the right attitude)

Traditional methods of teaching produce positive results in procedural fluency, but negatively impact the other four (Yoshinobu & Jones, 2012). IBL can benefit students in all five categories, and does not necessarily impede content coverage. Despite the coverage issue, research has shown that IBL students learn enough to succeed at least as well as their non-IBL peers in subsequent courses (Kogan & Laursen, 2014). Techniques for dealing with the coverage issue in an IBL class will be provided in later sections (Table 1).

Best Practices in IBL Teaching

IBL courses can vary, but the main ingredients of an IBL mathematics course usually include some or all of the following: collaborative learning, guided lessons which encourage reasoning and analyzing, presentations, and minimal lecture. Many IBL instructors might emphasize another key ingredient – selling the idea of IBL to the students. This section will begin with a discussion of how to get student buy-in, and then will follow with ideas about managing participation, homework, in-class work, assessment, portfolios, and exams.

Table 1. Some IBL Effects on Mathematical Proficiency.

1. Conceptual understanding	Constructing new examples, conjectures, or proofs leads to a deeper understanding
2. Procedural fluency	Spending class time solving problems or going over student work provides procedural understanding
3. Strategic competence	Struggling through problems without previously seeing similar examples allows students to learn problem-solving skills
4. Adaptive reasoning	Guided questions can lead students to think about what previous material will help them for a new problem
5. Productive disposition	The success of completing hard problems leads to pride and a better understanding of what mathematics involves as a discipline

Selling the Idea

Making sure that students are willing to engage in IBL is critical to having a successful semester. Mathematics students in particular have certain expectations of how the material will, and should be conveyed. They likely have only experienced lecture-based mathematics instruction. Many students feel math anxiety before they even step into the classroom (Dew et al., 1984; Hekimoglu & KittreII, 2010). Switching the teaching and learning style could add to their anxiety, especially when that style results in more effort and struggle on the part of the student. Even students who excel at mathematics and are comfortable with the subject may push back against IBL because, based on previous experience, they learned effectively through lecture. Spending time on student buy-in can help convince students that the skills learned through IBL, from communication to problem solving, will be beneficial in future classes and careers.

To sell the idea of IBL, explain what inquiry-based learning means to you, how it will be implemented, and why it is useful. A statement in the syllabus, which can be expounded upon on the first day, is a good way to set the tone for the semester. For example, in the syllabus you could write that IBL develops collaboration and communication skills. Then motivate the need for those skills by referencing the National Association of Colleges and Employers (NACE) compilation of "The 10 Skills Employers Most Want In 20-Something Employees," which lists the ability to work in a team as number one and the ability to communicate verbally as number four (Adams, 2013). Also, describe the importance of problem-solving and critical thinking skills. These attributes are apparent in the list of 10 skills from the NACE (Adams, 2013).

Another way to validate and explain the method to students is by comparing it to being on a sports team or learning an instrument. Many of your students have probably had experiences with being athletes or musicians, so you can connect with them through appropriate comparisons. Do soccer players learn to play solely by watching their coach do it? No, they practice, usually with their team. Does a trumpet player become better by watching someone else play the trumpet for an hour? Probably not. Then why would you learn mathematics by watching a teacher lecture on it? IBL gives students a chance to practice mathematics with their teacher and teammates in the room. It makes mathematics an active undertaking.

A critical aspect of selling the idea of IBL is to remind students throughout the semester of why you are using it. Even if you explain it the first day, it is helpful to reinforce the idea many times later. At the end of the course, students might have forgotten why they agreed it was beneficial months earlier.

A mid-semester survey to gage their thoughts or get suggestions can be a good opportunity to revisit those reasons and resolve complaints. At the end of the semester, initiate a final discussion about why IBL was used in the course, and how it may have helped them. One could give another survey asking about pre- and post-perceptions of group work, presentations, and other learning strategies. Another useful activity is asking current students to write a letter to future students, giving advice or explaining the benefits of IBL. If you put course objectives in your syllabus which include computation and conceptual goals, you can tie in the conceptual targets to what was completed in the IBL classroom. Anything that reminds them of the thoughtfulness behind choosing to teach with IBL can be helpful.

Participation and Presentations

Student participation in an IBL mathematics course can be implemented in more than one way. Some instructors prefer to adhere closely to Moore's method, where the majority of class time is spent on presentations of work done before class. Other IBL teachers devote a significant portion of the meeting period to group work, along with presentations. There are advantages to both approaches. Regardless, presentations often play a prominent role. These usually take the form of a student going to the board, writing down previously prepared work, and then explaining that work to the class. As the facilitator, you can choose to chime in to correct or refine board work, or simply ask leading questions until a student notices a mistake that you see.

Requiring students to attempt work individually before going over it in class can work well in proof-based courses. Students in such classes are usually motivated and interested in the material, since they are likely in a course for their major. They gain proof-writing skills and proficiency by struggling through problems on their own. Once in class, they can polish their work with the help of peers and the instructor. Teachers using this method often modify Moore's approach by decreasing the level of competition and increasing teacher involvement. Also, some instructors choose to require collaboration on the pre-class work by assigning groups that meet outside of class. In class, however, a single member of the group presents the solution.

Group work, in conjunction with presentations, can be implemented in both lower and upper level courses. Lower level mathematics courses are often taken by non-mathematics majors fulfilling a general education or major requirement. Allowing such students, who frequently have mathematics anxiety, to collaborate provides support for tackling difficult problems or concepts. Additionally, if they are working in the classroom then the teacher can provide guidance by checking on groups and answering questions (sometimes answering with another question), an opportunity not often present in traditional mathematics courses. Upper level students can benefit from group work when they get the opportunity to see how their peers begin problems or proofs, as opposed to seeing a finished product.

When class time is used for group work, students can present on this collaborative work or other homework done outside of class. For the former, if most groups finish their assigned tasks then you can transition straight into presentations during the same class period. If only a minority of groups finish, then either assign the rest for homework and start with presentations in the next class, or pick up with the group work again then.

To ensure that the participation component of the class is taken seriously, it should be a portion of the grade. Of course, then there must be a system of keeping track of participation and/or presentations.

Suggestions for monitoring presentations and participation:

- Write students' names on note cards and randomly choose a card. Then write the date of the presentation on the card, along with any comments about the presentation.
- Allow students to volunteer on a calendar or chart in a shared document; this way students can mark the problems they are ready to present before class begins. Then check the chart and choose presenters based on who is behind on their number for the semester.
- Allow pairs of students to go to the board, with one student as the main presenter and the other as back-up support. Having multiple students present a problem could be especially useful for large classes, or for introverted students.
- Collect the problems that were presented, but allow colored corrections where students use colored markers or pens to correct their work during presentations. The instructor then knows what the student did on their own.
- If grading participation, such as how many questions or answers a student provides, make a seating chart and mark down the seat of the

speaker. This is helpful when you do not yet know all of the students' names.

Non-Traditional Use of Traditional Materials

IBL teachers incorporate components used in traditional lecture-based courses, such as homework and lecture notes, in innovative ways. For example, if students are collaborating in class, homework might be assigned as individual work to ensure students practice problems on their own. If class time is spent on presentations, homework might be the collaboration phase. Since there is a minimal amount of lecture, IBL students do not take as many lecture notes. They may have a textbook, or they might be given a script of definitions, axioms, etc., to work from. Such changes lead to a mathematics class that significantly differs from a traditional course because there is more student involvement.

Although textbooks can be used in IBL classes, some instructors prefer to have more control over when and how material is given to their students. For instance, if you want students to prove a certain theorem, and the text has that theorem, then they could potentially just look at it in the text. As a result, many IBL teachers create or adopt their own course text, or script.[3] A script differs from the usual textbook because a lot of explanations and solutions are left out, with the intention that the students fill in the gaps. If group work is emphasized, the teacher could hand out daily worksheets. Even a worksheet can be implemented in a new way. Instead of the common practice of asking questions that are almost identical to previously seen examples which were worked out by the teacher, the worksheet should have more conceptual, new, or challenging problems. Teaching a class using IBL can be a time-consuming process for instructors, since they may be creating new scripts or worksheets. Having a textbook to reference and use for homework problems alleviates some of the problem. For classes that have a formal textbook, professors can emphasize that students should only refer to the book after material has been discussed in class.

To encourage inquiry and conceptual learning should be the main objective of the material given to students, whether in a book, script, or worksheet format. For example, if students are learning about probability formulas, then ask them to generate real-world examples of when the formulas would be used. For another illustration, consider the comparison test or theorem for infinite series in calculus. The idea of the theorem is that if a value is larger than infinity, then that value must also be infinite.

Or, if the number is between 0 and a finite number, then that number must be finite. By providing a few illuminating examples and then asking students to generalize the idea, the instructor can guide students to develop the theorem, likely leading to a greater understanding than if it had been presented to them. Many mathematical concepts, regardless of the course, can be taught in this way. In lower level courses, however, the material provided to students might include more given information. Students can still be asked to try unfamiliar problems or create examples using that new information.

Summary of previous suggestions:

- Using a textbook can be helpful, especially if you are teaching the class as IBL for the first time.
- If you have more time for lesson planning, try to create your own text/script of the exact material that you want students to see.
- Ask conceptual questions! If possible, guide students to develop definitions, formulas, theorems, proofs, or examples on their own.
- Do not simply give worksheets with repetitive problems, or problems similar to previously seen examples.
- For group work classes, assign "individual" homework problems so that students must do some work without the support of a group.[4]

Additional suggestions:

- Check the *Journal of Inquiry-Based Learning in Mathematics* (*JIBLM*) at jiblm.org for peer-reviewed course materials.
- For groups, assign especially challenging problems. It is easier to solve a hard problem when you have others to suggest ideas and bounce your own ideas off of.

Beyond Traditional Testing

Often IBL courses involve a lot of talking. However, in contrast to lecture-based courses, the speakers are students. Whether through group work or presentations, each student has plenty of opportunity to explain mathematics, and should be expected to do so. Learning mathematics and learning to communicate mathematics go hand in hand. In traditional mathematics courses, when students are spending the majority of their class time listening and writing, it makes sense to test them mainly through written exams. In an IBL course, oral exams can have a place as a complement to written exams. Written tests are still useful, especially to be able to test larger amounts of material. If oral exams are done individually with students, they can be very time-consuming. If the instructor is able to commit that time, though, there are several benefits to oral exams.

Oral exams emphasize the importance of communication skills and give the instructor a real-time view of how a student approaches a problem and solves it. It is easier to see what topics or ideas cause struggles, as the student talks through his or her thought process. For example, a student may be able to use a definition or theorem correctly without truly understanding the concept behind it. Such lapses of knowledge can be drawn out during oral questioning.

While intimidating at first, oral exams can actually boost student confidence. Success after such an endeavor can increase the students' faith in their abilities. To improve students' probability of success, allow them to ask for help or hints for a certain amount of points. This way, they should be able to finish the problem. You can make the exam more like a conversation, as opposed to just drilling them with questions or saying nothing as the student talks. Oral exams can ease the minds of students who suffered from test anxiety, once they realized that they have the help of the instructor to get through the oral problem. The oral exam takes stress off of the written exam, if the two portions each account for part of the test grade (Capaldi, 2014).

Suggestions for oral exams:

- Schedule more time than you expect to actually need for the exams. You might be surprised about how long students take to answer the question(s). In proof classes, usually one proof is enough for a 10–20 minute individual oral exam.
- Ask leading questions if the student is stuck, like "What is the definition of blah-blah again?," if that is the definition they need to use to move forward.
- If the student is moving in the wrong direction, do not let him or her proceed too far. Ask questions along the lines of "How do you know you can do that?," or tell him or her that you will give them a hint for points.
- To decrease anxiety, give a list of previously seen problems for students to use to study for the first exam. Then, for the second (and/or third) exam, give them a new problem during the exam.
  - If using a list of problems, you can tailor the choice of which ones you pick to suit the student's ability level. For high-level students,

challenge them with a harder problem. You can grade it with more leniency.

  - If you give a new problem, and trust your students not to share, then give everyone the same problem.

ENCOURAGING STUDENT REFLECTION

In an IBL class, students create a lot of impressive examples and proofs that are a source of pride. Portfolios and student-created textbooks provide a means for students to collect and demonstrate that knowledge gained throughout the semester. If students are required to revise their work before putting it in the portfolio or book, then they have additional motivation to rethink mistakes and review material.

Students often do not realize how much they have accomplished by the end of the semester, and that effort is nicely demonstrated through an organized collection of their work. Portfolios can contain proofs or computational solutions that were previously worked on in class or for homework. Even when students see correct solutions on the board during presentations, their own notes remain flawed. By checking that work in the portfolio, the instructor can point out any remaining mistakes and allow the students' work to be revised before the final submission of the portfolio. Similarly, students rarely turn in perfect homework the first time around. If the homework solutions are included in the portfolio, then they are greatly improved after another round of correction. The major downside to portfolios is that checking and grading them can be a large time commitment for the instructor. For large classes, they may not be practical.

An alternative to individual student portfolios is a student-created textbook. This medium allows the entire class to contribute to one document, making the teacher's grading much more manageable. There are relatively easy ways to administer a collaborative text, such as wikibook.org or a shared Google document. Since every student adds to this one text, they can be given rotating roles, such as being in charge of certain definitions, examples, or theorems. Some students could be given the tasks of checking or revising what was posted. Then the instructor simply checks to make sure each student is fulfilling his or her duty.

Suggestions for portfolios:

- Because of the time commitment, only review or grade portfolios two to three times in the semester.

- Ask students to leave previously reviewed work in a separate part of the portfolio, so you can see what comments or mistakes you marked before, and check whether they were fixed.
- Other problems and benefits of portfolios are discussed in an article about teaching abstract algebra (Capaldi, 2014).

Suggestions for student-created textbooks:

- Online educational platforms like Blackboard.edu provide a wiki option.
- A website called writelatex.com allows users to contribute to a shared file in LaTeX, a commonly used program for typing mathematical documents.
- A single class textbook is good for large classes. Just give students rotating roles.
- Students end up with a nice study guide which they have produced.
- Reviewing the text allows students to evaluate each other's work.

CHAPTER SUMMARY

Inquiry-based learning in mathematics took root over 100 years ago in the teaching of R.L. Moore, and has developed into an effective and widespread method. Research suggests that active learning increases conceptual understanding by mathematics students, and that IBL in particular benefits low-achieving students as well as women in mathematics. IBL naturally leads students to mathematical proficiency and instills in them a sense of the creativity involved in doing mathematics.

Depending on their specific needs and teaching styles, instructors incorporate IBL differently into their mathematics curriculum. Often, this implementation involves group work and presentations; suggestions for these activities were provided in this chapter. Additionally, the author recommends consideration of oral exams, which emphasize the importance of communication skills and sometimes demonstrate students' understanding of mathematics more clearly than written exams. Finally, student portfolios or student-created textbooks encourage review and reflection of the semester's work.

IBL can be used in many ways and in any mathematics course. Furthermore, the ideas of this chapter could be applied to non-mathematics courses. Large or small, upper or lower level, a course can be altered to include inquiry-based activities. Even if only portions or specific topics are taught in an inquiry-based manner, students can still benefit from experiencing mathematics in an engaging way.

NOTES

1. Moore famously produced 50 doctoral students, many of whom went on to have their own great achievements in mathematics (Fitzpatrick). Their later success may have been due to such early exposure to proving and discovering on their own (Frantz, 1983).

2. There is an annual "Legacy of R.L. Moore" conference (http://legacyrlmoore.org/events.html) that provides an opportunity for people to share their ideas on non-traditional methods of teaching mathematics, mostly centering around inquiry-based learning.

3. For a good example of a script, see Ron Taylor's notes for an Introduction to Proofs course at http://www.jiblm.org/downloads/jiblmjournal/V070404/V070404.pdf (Taylor, 2007).

4. The writer explains personal experiences with the issue of homework collaboration in an article on non-traditional methods of teaching abstract algebra (Capaldi, 2014).

REFERENCES

Adams, S. (2013). The 10 skills employers most want in 20-something employees. Retrieved from http://www.forbes.com/sites/susanadams/2013/10/11/the-10-skills-employers-most-want-in-20-something-employees/. Accessed on July 14, 2014.

Capaldi, M. (2014). Non-traditional methods of teaching abstract algebra. *Primus: Problems, Resources, and Issues in Mathematics Undergraduate Studies*, *24*(1), 12–24.

Dancis, J., & Davidson, N. (1970). The texas method and the small group discovery method. Retrieved from http://legacyrlmoore.org/reference/dancis_davidson.html

Dew, K. H., Galassi, J. P., & Galassi, M. D. (1984). Math anxiety: Relations with situational test anxiety, performance, physiological arousal, and math avoidance behavior. *Journal of Counseling Psychology*, *31*, 580–583.

Epstein, J. (2013). The calculus concept inventory-measurement of the effect of teaching methodology in mathematics. *Notices of the AMS*, *60*(8), 1018–1026.

Frantz, J. B. (1983). The moore method. *The Forty Acres Follies* (pp. 111–122). Austin, TX: Texas Monthly Press.

Freeman, S., Eddy, S. L., McDonough, M., Smith, M. K., Okoroafor, N., Jordt, H., & Wenderoth, M. P. (2014). Active learning increases student performance in science, engineering, and mathematics. *Proceedings of the National Academy of Sciences*, *111*(23), 8319–8320.

Good, C., Rattan, A., & Dweck, C. S. (2012). Why do women opt out? Sense of belonging and women's representation in mathematics. *Journal of Personality and Social Psychology*, *102*(4), 700–717.

Hekimoglu, S., & Kittrell, E. (2010). Challenging students' beliefs about mathematics: The use of documentary to alter perceptions of efficacy. *Primus: Problems, Resources, and Issues in Mathematics Undergraduate Studies*, *20*(4), 299–331.

Jones, F. B. (1977). The Moore method. *American Math Monthly*, *84*, 273–277.

Kogan, M., & Laursen, S. (2014). Assessing long-term effects of inquiry-based learning: A case study from college mathematics. *Innovative Higher Education*, *39*(3), 183–199.

Kwon, O. N., Rasmussen, C., & Allen, K. (2005). Students' retention of mathematical knowledge and skills in differential equations. *School Science and Mathematics*, *105*, 1–13.

Laursen, S., Hassi, M., Kogan, M., Hunter, A., & Weston, T. (2011). *Evaluation of the IBL mathematics project: Student and instructor outcomes of inquiry-based learning in college mathematics*. Boulder, CO: University of Colorado. Retrieved from http://www.colorado.edu/eer/research/steminquiry.html

National Science Foundation. (1996). Shaping the future: New expectations for undergraduate education in science, mathematics, engineering, and technology, advisory committee to the NSF directorate for education and human resources. Retrieved from www.nsf.gov (file nsf96139)

Schoenfeld, A. H. (1988). When good teaching leads to bad results: The disasters of "Well-Taught" mathematics courses. *Educational Psychologist*, *23*(2), 145.

Seymour, E., & Hewitt, N. M. (1997). *Talking about leaving: Why undergraduates leave the sciences*. Boulder, CO: Westview Press.

Springer, L., Stanne, M. E., & Donovan, S. (1999). Measuring the success of small-group learning in college level SMET teaching: A meta-analysis. *Review of Educational Research*, *69*, 21–51.

Taylor, R. (2007). Introduction to proof. *Journal of Inquiry-Based Learning in Mathematics*, (4). Retrieved from http://www.jiblm.org/downloads/dlitem.aspx?id=56&category=jiblmjournal

Ward, B. B., Campbell, S. R., Goodloe, M. R., Miller, A. J., Kleja, K. M., Kombe, E. M., & Torres, R. E. (2010). Assessing a mathematical inquiry course: Do students gain an appreciation for mathematics? *Primus: Problems, Resources, and Issues in Mathematics Undergraduate Studies*, *20*(3), 183–203.

Yoshinobu, S., & Jones, M. G. (2012). The coverage issue. *PRIMUS*, *22*(4), 303–316. doi:10.1080/10511970.2010.507622

ENHANCING INQUIRY-BASED LEARNING ENVIRONMENTS WITH THE POWER OF PROBLEM-BASED LEARNING TO TEACH 21ST CENTURY LEARNING AND SKILLS

Henk Huijser, Megan Y. C. A. Kek and Ruth Terwijn

ABSTRACT

This chapter provides an outline of how the essential elements of problem-based learning (PBL) can be adapted to enhance inquiry-based learning environments and in the process teach 21st century skills. It uses a case study of a first-year nursing course at a regional Australian university to show how essential PBL elements can be adapted in an 'ePBL' context, following five ePBL steps. Overall, it is argued that a carefully mapped outset of learning outcomes and PBL problems designed as inquiry-based activities provide a 'liquid learning' environment that will

Inquiry-Based Learning for Science, Technology, Engineering, and Math (STEM) Programs: A Conceptual and Practical Resource for Educators
Innovations in Higher Education Teaching and Learning, Volume 4, 301–320

ISSN: 2055-3641/doi:10.1108/S2055-364120150000004017

ultimately prepare confident graduates who will be able to take full advantage of the 21st century learning and professional contexts in which they find themselves.

INTRODUCTION

Inquiry-based learning is a broad umbrella term to describe approaches to learning that are driven by a process of inquiry, including project-based learning, discovery learning, case-based learning, and problem-based learning (PBL) (Kahn & O'Rourke, 2005; Prince & Felder, 2006). In general, we perceive inquiry-based learning as a type of learning that is focused on an individual student's conceptual transformation, and an approach that is based on posing questions, gathering and analysing data, and constructing evidence-based arguments. By contrast, PBL leans more towards social constructivism where individuals are always situated in, and constituted by, the social and cultural contexts in which they find themselves. Though this difference may be seen as quite subtle, the implications for designing a truly student-centred curriculum and learning environment that motivates students to be curious, explore, and investigate and simultaneously learn disciplinary content and strategic skills such as critical thinking, reasoning, team work, and self-directed learning, are rather large. These skills, which were being developed as part of a PBL approach nearly 50 years ago, remain critical skills today, in an increasingly interconnected world where interactions between individuals and their technology-infused environments remain an ever-changing constant, a world where learning environment are increasingly 'liquid' (Savin-Baden, 2014). In this chapter, we explore ways in which PBL design principles and teaching practices can enhance inquiry-based teaching and learning environments in higher education across different disciplines and cultures, and thereby equip students with the skills they need to succeed into the future.

This chapter offers an exploration of the application and power of PBL and its founding principles (Barrows & Tamblyn, 1980) when designing a student-centred inquiry-based curriculum that aligns with 21st century learning outcomes. Contemporary learning takes place in an environment with 'the rich resources provided by the digital information network – where the context in which learning happens, the boundaries that define it, and the students, teachers and information within it all co-exist and shape each other in a mutually reinforcing way' (Thomas & Brown, 2011, p. 35). We believe that the power of PBL, when applied appropriately and creatively, can be used to revitalise an inquiry-based learning environment, as the original intentions of PBL are applicable across many disciplines and cultures, which will be outlined in-depth in the section 'Essentials of PBL'. We present the essentials of PBL to develop 21st century skills that can then serve as a guideline to curriculum design and that make use of Biggs' (Biggs & Tang, 2011) constructive alignment concept developed in the 1990s. Finally, we use a limited case study of a nursing programme at an Australian university to show how these ideas could be applied in the context of Science, Technology, Engineering, and Mathematics (STEM) education.

NEW LEARNING CULTURES IN AN INTERCONNECTED 21ST CENTURY WORLD

Profound changes have taken place in working environments globally, including in higher education, under the influence of what some call the 'digital revolution' (Davies, 2012). This revolution has not only changed the ways in which learners access information (and by extension education in general), but also the role of traditional educational institutions, and most importantly, the ways in which such institutions might prepare learners to function effectively in this changed environment (Bradwell, 2009). Deakin-Crick (2007) contributes to this discussion by describing a need for a 'paradigm shift towards a relational and transformative model of learning, in which the creation of interdependent communities of intentional learners provides a basis for the integration of 'traditional academic' skills and outcomes with the learning dispositions, values and attitudes necessary to meet the demands of the emerging 'networked society'' (p. 137). This paradigm shift is by now well underway and it is creating a fast evolving fluid environment which currently includes Massive Open Online Courses (MOOCs), flipped classrooms (classroom used for discussion; lecture material to be read or viewed outside of class time), dynamic assessment (interactive assessment) and event-based learning (specifically designed events to make the learning memorable) (Sharples et al., 2014). Many of these phenomena are surrounded by hype and need to be approached with caution, but they provide insight into evolving learning environments that require different and flexible approaches to learning and teaching in higher education. The key focus is on 'intentional learners' (Deakin-Crick, 2007), which suggests lifelong learners who are metacognitively equipped to identify what and when they need to learn, and how to do so most effectively. In other words, these are learners who take control of their own learning, who have an awareness of their own strengths and weaknesses, and who use this awareness to both learn and grow their own learning capabilities and those of the communities they are part of or participate in.

When considering these changes, obvious questions arise of how to respond to them in higher education settings, particularly in terms of curriculum design and learning and teaching approaches. We suggest that contemporary learning environments demand of us to not just keep up with the rapid rate of changes but view 'learning in terms of an environment – combined with the rich resources provided by the digital information network – where the context in which learning happens, the boundaries that define it, and the students, teachers and information within it all co-exist and shape each other in a mutually reinforcing way' (Thomas & Brown, 2011, p. 35). Savin-Baden (2014) has talked in this respect about 'liquid learning' – 'the sense that learning and knowledge are always on the move, so that faculty design curricula in ways that introduce questions about practices and understandings of knowledge within and beyond disciplinary areas' (p. 210). She further suggests that 'PBL liquid learning curricula need to be characterised by emancipation, reflexivity, and flexibility so that knowledge and knowledge boundaries are seen as contestable and always on the move' (Savin-Baden, 2014, p. 210). She begins to define what kinds of curricula are needed in what is essentially an increasingly 'liquid' learning context. However, while this may imply profoundly unstable conditions, it is also a space of potentially boundless opportunities for those who know how to engage in this space and leverage the benefits. So how do we develop learning communities that are equipped to take full advantage of contemporary learning spaces and opportunities? Enter inquiry-based learning.

INQUIRY-BASED LEARNING (IBL)

Inquiry-based learning (IBL) is an overarching term to describe a number of student-centred approaches of teaching that promote student learning by engaging and involving students in the process of research or search for new knowledge to answer a question or problem or task (Deignan, 2009; Hutchings, 2007; Justice et al., 2007; Kahn & O'Rourke, 2005; Prince & Felder, 2006; Spronken-Smith & Walker, 2010). IBL is commonly considered to encompass PBL, project-based learning, discovery learning, and case-based learning (Kahn & O'Rourke, 2005; Prince & Felder, 2006). As Manista and Gillespie (2011) argue, '[IBL] is not simply one methodology that resists any blending. On the contrary, it is relatively fluid and can accommodate a variety of teaching practices and learning styles' (p. 88), which is not to suggest that it has no boundaries. The idea of fluidity echoes Savin-Baden's (2014) 'liquid learning', which is not a coincidence, as IBL allows for a level of unpredictability and flexibility, which in turn is central to student-centred and constructivist approaches. As Manista and Gillespie (2011) note, 'the idea is to get the students involved with the material directly, as opposed to offering them a supposed authoritarian interpretation' (Manista & Gillespie, 2011, p. 87). We can take that even a step further by involving students in selecting the material, rather than having teachers choose for them.

In IBL, students generate facts, rules, or procedures, at which point they are either presented with information or facilitated by their teachers to discover it for themselves. As such, IBL learning and teaching are also known as inductive learning and teaching when compared to the more traditional, teacher-directed forms of learning and teaching that are deductive (Prince & Felder, 2007, 2006). Students in IBL can experience different inquiry types which can range from structured, guided or open (Prince & Felder, 2006), which again implies a level of fluidity. Structured inquiry is an inquiry approach where students are provided with a question or problem and an outline is provided on how to solve it. In guided inquiry, students are assisted in finding a solution, whereas in open inquiry, students formulate the problems or questions themselves, even if this is obviously a sliding scale. In work-based contexts beyond their university studies, students are likely to encounter situations that require any one of these types of inquiry, so essentially they need to be able to assess appropriate responses and forms of inquiry and then apply them.

At the core of IBL approaches lies the open-ended or close-ended question/s that students answer. The sole function of the questions is to stimulate or motivate students into embarking on a process of seeking answers to the questions posed by their teachers (Hutchings, 2007). Levy and Petrulis (2012) have conceptualised a continuum of inquiry, ranging from inquiry for learning, which is about exploring existing knowledge, to inquiry for knowledge building. Their conceptual proposition is an attempt to move IBL beyond the notion of an active learning paradigm to potentially undertaking real research, moving the 'learning paradigm' or inquiry for learning to a 'discovery paradigm' or inquiry for knowledge building in undergraduate education (Levy & Petrulis, 2012, p. 97). A discovery paradigm would take it into the realm of 'liquid learning'.

The theoretical support for IBL is underpinned by research into student learning with roots in a broad constructivist approach. The key tenet of constructivism is that an individual student must actively construct knowledge and skills. The two main forms of constructivism, Piaget's (2006) theory of cognitive development and Vygotsky's (1978) social constructivism, apply to IBL, as does Wood, Bruner, and Ross's (1976) explanation of instructional scaffolding, which focuses on the role of the tutor. Piaget's cognitive constructivism proposed that individuals must construct their own knowledge and they build it through their experience in the world. By contrast, in social constructivism, social context and interactions within it are the influencing basis for learning. Vygotsky's (1978) concept of the zone of proximal development posits that individuals can, with the support of a more experienced peer, master ideas and concepts that they would not have been able to on their own. Grounded in these underlying constructivist principles, IBL has a number of key dimensions.

Spronken-Smith and Walker (2010) outline the key focus of an IBL approach as follows:

- Learning is stimulated by inquiry, mostly driven by questions or problems;
- Learning is based on a process of building knowledge and new understandings;
- It is an active approach to learning which involves learning by doing;
- The role of the teacher is to act as a facilitator; and
- Students increasingly take responsibility for their learning and being self-directed (p. 726).

This key focus could easily be read as the focus of a PBL approach as well.

THE POWER OF PBL FOR ENHANCING IBL

IBL is commonly mistaken for another pedagogical approach known as PBL or the terms are used interchangeably. Both IBL and PBL might 'share the same philosophical orientations and educational intentions' (Mantzoukas, 2007, p. 245), but the key essentials and educational intentions of PBL might enhance IBL environments. Conversely, IBL may be preferred to PBL because PBL is sometimes seen as too rigid and not flexible enough (Manista & Gillespie, 2011). However, we use a conceptualisation of PBL that is much closer to the idea of 'liquid learning'.

The power of PBL lies in its agility, and in its ability to blur the boundaries between higher education and the world beyond. PBL should be designed to trigger multiple interactions between educational environments and wider society so that students are ultimately able to seamlessly move into their chosen profession and/or vocation. Knowledge is generated and learning occurs through interactions in small groups, working together to solve ill-structured, complex, or open problems. Students not only acquire declarative knowledge enmeshed in the problems but also strategic functioning skills, such as problem-solving, collaboration, self-directed learning, and critical thinking/reasoning skills, and the ability to apply them in ever-changing contexts. Such skill development is aligned to the original conceptualisation of PBL's outcomes in medical education (Barrows & Tamblyn, 1980), and yet transferable to a 'liquid learning' context. PBL has the potential to enhance IBL environments through its agility and adaptability to the complexities of contemporary learning contexts. Moreover, it has the power to provide students with both disciplinary knowledge and strategic functioning skills needed to succeed and thrive in 21st century environments. The focus should be on the underlying principles and envisaged learning outcomes, rather than on distinctions between problem-based, project-based, or IBL. In a 'liquid learning' context, students and teachers should be able to move seamlessly between these approaches, and apply the most appropriate approach in any given situation.

Since its beginning in the 1960s, PBL has expanded to almost all disciplines and fields, levels of studies, and countries on five continents (Hung & Loyens, 2012). PBL has roots in constructivism and situated learning theories (Marra, Jonassen, Palmer, & Luft, 2014). Jonassen's (1991) concept of constructivism, which includes five pillars that together combine the elements of knowledge, meaning making and learning, forms one of the theoretical underpinnings of PBL. The five pillars of constructivism are: (1) knowledge is a product of students' sense making through their interactions with the environment; (2) knowledge is a reality to its beholder, and thus knowledge is not simply an external entity that exists in the world to be passed on to others; (3) meaning and thinking are distributed in an environment of cultures, people, and tools, and thus social interactions and interdependence in group settings or communities are important in PBL; (4) relevant contexts where knowledge is acquired and skills are applied are important for sense making and (5) knowledge construction (learning) is stimulated by a problem or need to know.

The other theoretical underpinning of PBL has roots in situated learning. Situated learning theorists have argued that meaningful and lasting learning occurs best when the learning process mirrors as closely as possible the environment where the learning will be applied (Brown, Collins, & Duguid, 1989), which is also directly relevant to PBL in a contemporary context. Hung (2002) argues that authentic contexts are important in situated learning environments because knowledge can only be generated through multiple exchanges and interactions between the minds of individuals, and the environments in which they are situated.

ESSENTIALS OF PBL

Learning Is Action Oriented, In-Context, and Just-In-Time

Boud and Feletti (1998) argue that 'problem-based learning is grounded in the belief that learning is most effective when students are actively involved and learn in the context in which knowledge is to be used' (p. 4). Action and situated learning afford students the ability to solve complex problems and to significantly increase the quality of learning. In PBL, students habitually seek, analyse, synthesise and apply information, with the guidance of the teacher, to resolve the problems presented to them at the outset. Authentic problems should be used so as to further fuel students' interest and motivation and to trigger inquiry. This means that problems will often be interdisciplinary and multifaceted, inherently triggering liquid learning.

Learning Needs Explicit Modelling and Scaffolding with Fading

Successful learning happens when there is a scaffolding process with fading. Scaffolding is a term first introduced by Wood et al. (1976) and refers to those instances when appropriate assistance and support are provided, and learners can attain intended goals or engage in a practice that would otherwise be beyond their skill level. The notion of scaffolding is also consistent with Vygotsky's (1978) conception of the zone of proximal development (ZPD). However, scaffolding *without* fading (Pea, 2004) will not allow students to perform autonomously. Therefore, it is essential that teachers avoid making the students dependent on them, but instead use the model-scaffold-fade approach to teaching (Kek & Huijser, 2011). That is, the teachers pose questions through modelling and scaffolding, to gradually but consciously remove the modelling (allowing the students to adopt the questioning role instead of the teachers) and scaffolds (support mechanisms for students) and to 'fade' themselves out of the picture, thereby ensuring that students become independent learners.

Learning Occurs in Groups, with Others and Peers

In an increasingly digitally interconnected world, knowledge is constructed collectively, physically and virtually. It is through conversations with others about what each individual knows about the problem that learning occurs (Pask, 1976), and knowledge is shared and built (Hmelo-Silver & Barrows, 2008; Scardamalia & Bereiter, 2006). Collaborative learning in groups should therefore be embedded in the curriculum.

Learning Needs to Be Visible

An integral aspect of learning in groups through conversations is the need to externalise understanding (Pask, 1976). To be able to engage in a productive conversation, all members of the group, including the teacher, need to have a common understanding of the problem. Barrows' reasoning structure (Wee & Kek, 2002) – facts (identify the facts), ideas (generate or explore possible ideas or hypotheses), learning issues (identify what you need to know to manage or solve the problem), and action plan (articulate an action plan to seek, evaluate, synthesise and apply information to the problem), also known as the FILA table – is a tool to scaffold learning by making thinking and reasoning processes visible and explicit to all parties in the PBL session, including students and teachers. FILA provides a structure for students, facilitated by teachers, to plan the best way to address a specific problem, as it forces students to ask questions about what they already know, and in turn what they need to learn.

Learning Requires Closing the Loop on Learning through Effective Feedback and Reflections, Self- and Peer Assessment

It is important to close the loop on learning with reference to how every student in the group performs in PBL. It is also an important element in developing metacognitive skills. Providing and receiving feedback are another essential element of PBL. It is expected that the reflections form a part of structured learning activities or regular feedback sessions, usually at the end of each problem. The goal is to nurture the development of greater self-awareness of one's own strengths and weaknesses, an important skill of critical thinkers (Halpern, 1998), and, in particular, an essential skill in contemporary learning and professional contexts (Rotherham & Willingham, 2010). Closing the loop thus facilitates the development of reflective practitioners (Schon, 1983).

Learning Is Relational, Requiring Creation of Conditions for Deep Learning

Learning is not stable. Phenomenographic studies and research following the tradition of Marton and Saljo (1976), which focus on qualitative differences in individual approaches to learning, have established that a surface approach to learning is not necessarily inherent to students. Student approaches to learning studies using quantitative methods also have found students' learning to be associated with their perceptions of the learning contexts (Biggs, 2001). Biggs (1999) argued that only when the learning context involves a curriculum that is constructively aligned, would students be 'entrapped in [a] web of consistency, optimising the likelihood that they will engage the appropriate learning activities' (p. 5). While this may appear out of step with liquid learning, it is far from it, and it rather depends on what the envisaged learning outcomes are. What it means, however, is that to successfully implement PBL, the intentions of developing higher order thinking skills such as critical thinking, problem-solving, reasoning, and self-directed learning while learning discipline knowledge or subject matter, must be *intentionally* aligned with the essentials of PBL.

Learning Is Most Effective in a Constructively Aligned Environment

According to Biggs and Tang (2011), constructive alignment:

> … extends in a practical way Shuell's statement that 'what the student does is actually more important in determining what is learned than what the teacher does'. The intended outcomes specify the activity that students should engage in if they are to achieve the intended outcome as well as the content the activity refers to. The teacher's task is to set up the learning environment that encourages the student to perform those learning activities, and to assess student performances against the intended learning outcomes (p. 97).

Of course, the learning environment can consist of a combination of online and face-to-face components, and can be deliberately 'liquid' in nature. None of this necessarily negates constructive alignment, which has two aspects:

- The *constructive* aspect which refers to *what students do*, which is to make meaning or construct meaning through learning activities; and
- The *alignment* aspect which refers to *what teachers do*, which is to create the learning environment that supports the learning activities and processes appropriate to achieving the intended learning outcomes.

To these two aspects, a third might be added: evaluation of learning and intended outcomes. Evaluation is 'a process by which the effectiveness of educational interventions can be assessed' (Ascroft & Palacio, 1996, p. 93). The evaluative results are the quantitative analyses or evidence of learning quality and outcomes. They serve to answer questions from management and peers who might be skeptical about PBL in general, and about whether change is really an improvement (Paul, 2010). In addition, they provide ways to continuously improve PBL, which is particularly important in contemporary 'liquid learning' contexts. The constructive alignment idea infers that, if the educational goal is to prepare students for future learning so that they become adaptive and agile learners and experts, PBL as curriculum, learning environment, assessment, teaching and evaluation must be aligned in an interconnected and integrated teaching and learning system.

CASE STUDY: APPLICATION OF PBL ESSENTIALS IN A FOUNDATION NURSING COURSE

The implications of applying the essentials of PBL are demonstrated within the following case study. This is very much a context-driven process, which depends on where students are at on the one hand, and what the desired learning outcomes are on the other. This section showcases the online PBL learning design for 'responsible nursing practice', a first-year foundation nursing course at a regional university in Queensland, Australia. The course develops responsible nursing practice through the application of knowledge within person-centred care, ethical practice and therapeutic communications. The learning outcomes of the PBL tutorials are to acquire nursing knowledge and simultaneously develop students' critical thinking and clinical reasoning, self-directed learning and collaborative skills. PBL as a teaching approach was first adopted and introduced in its on-campus, face-to-face tutorials in 2010 (see Martyn, Terwijn, Kek, & Huijser, 2014). In 2012, in addition to the on-campus, face-to-face study mode, the undergraduate nursing programme was extended for the first time to distance education students who enrol to study fully online. Therefore, an online version of PBL tutorials and activities was developed and implemented to these distance education students. The educational goals, PBL teaching approach, assessments and learning process of the online PBL tutorial are similar to the on-campus PBL tutorial. Students in the two modes of study encounter three nursing problems in the 12-week semester.

The PBL nursing problems are centred on multidimensional issues that exist in the real world. The problem featured in this case study is of Jessica Smith, a 25-year old female presenting to A & E with urinary symptoms and a fever. Hospital notes provide factual information and artefacts to prompt learning issues and action plans to access appropriate and relevant information. The importance of history taking is highlighted with students as they develop questions about the patient's condition.

Group (or team) work is central to the success and learning outcomes associated with PBL and ePBL. Sharing the load of researching evidence-based information within nursing care underpins collaboration. Bringing large and varying amounts of information into classrooms and online platforms creates an in-depth approach to the knowledge gained and learned.

Once the information has been discussed and disseminated within the class, a mind or concept map is applied to organise and connect the learning for the nursing problem. A free online mapping software, *FreeMind*, was introduced to students. These concept maps are large but provide a visual representation of all information gathered for the nursing problem. Evolving concepts are then connected. Assessment of this course was in the form of a concept map using the earlier mentioned FILA table as the guiding framework (see Fig. 2).

The PBL essentials, discussed previously, are inextricably linked to the online PBLversion or the ePBL process (steps) as demonstrated in Fig. 1. The ePBL process is supported by the PBL essentials to achieve a constructively aligned PBL environment.

The following is the online ePBL process that students encounter:

ePBL Step 1: Asynchronous Group Discussion Forum

The problem is introduced to students in their group discussion forum before they meet their tutors in an online synchronous tutorial. The 'problem' may be a 'traditional' ill-structured PBL problem, but may also be a project brief or inquiry-based task, depending on the goals of instruction. At this stage, the students, on their own, would start thinking about the problem using the FILA table.

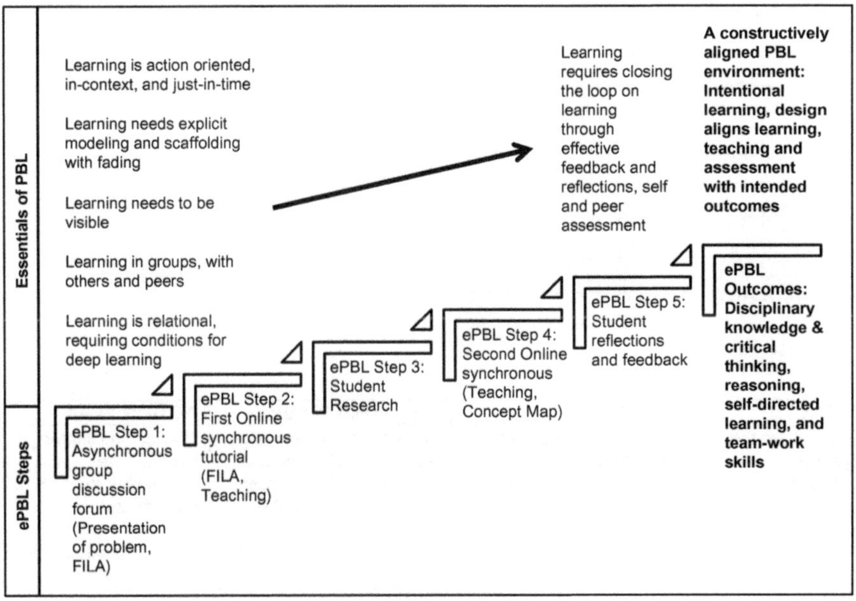

Fig. 1. Mapping ePBL Steps to PBL Essentials.

ePBL Step 2: First Online Synchronous Tutorial

Respective groups would meet their tutor for their online tutorial at the nominated time and day to discuss the problem. At this stage, artefacts are sometimes introduced along with the problems such as admission procedures, provisional diagnosis, doctor's orders, notes and vital signs documentation. In this first online synchronous tutorial, the tutor would facilitate the group learning, using the FILA table to discuss the problem as demonstrated in Fig. 2. The FILA is constructed using a wiki as a tool in the Moodle (Learning Management System) environment. Students use a colour coding system to demonstrate their contributions.

ePBL Step 3: Student Research

Students perform their research activities individually, while contributing to the group's research wiki in Fig. 3. The wiki provides the area where the group's research is curated for synthesis and elaboration during the second online and synchronous tutorial.

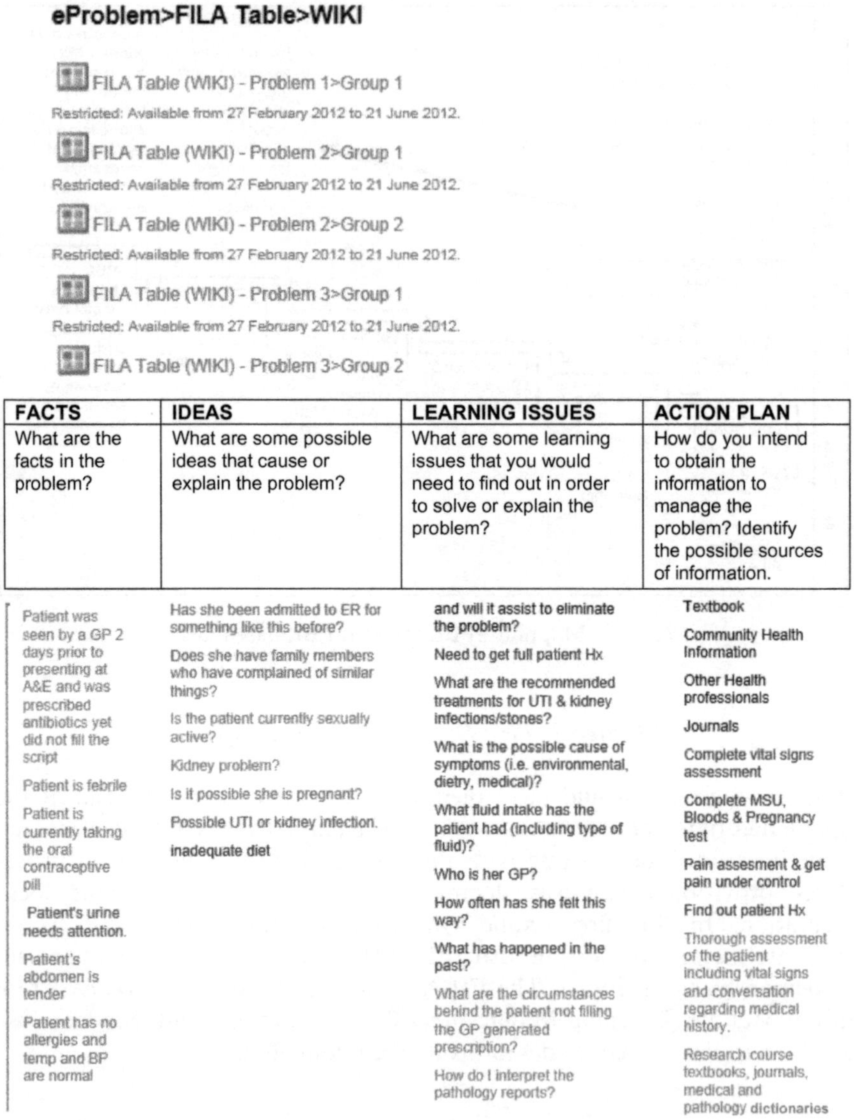

eProblem>FILA Table>WIKI

FILA Table (WIKI) - Problem 1>Group 1

Restricted: Available from 27 February 2012 to 21 June 2012.

FILA Table (WIKI) - Problem 2>Group 1

Restricted: Available from 27 February 2012 to 21 June 2012.

FILA Table (WIKI) - Problem 2>Group 2

Restricted: Available from 27 February 2012 to 21 June 2012.

FILA Table (WIKI) - Problem 3>Group 1

Restricted: Available from 27 February 2012 to 21 June 2012.

FILA Table (WIKI) - Problem 3>Group 2

FACTS	IDEAS	LEARNING ISSUES	ACTION PLAN
What are the facts in the problem?	What are some possible ideas that cause or explain the problem?	What are some learning issues that you would need to find out in order to solve or explain the problem?	How do you intend to obtain the information to manage the problem? Identify the possible sources of information.
Patient was seen by a GP 2 days prior to presenting at A&E and was prescribed antibiotics yet did not fill the script Patient is febrile Patient is currently taking the oral contraceptive pill Patient's urine needs attention. Patient's abdomen is tender Patient has no allergies and temp and BP are normal	Has she been admitted to ER for something like this before? Does she have family members who have complained of similar things? Is the patient currently sexually active? Kidney problem? Is it possible she is pregnant? Possible UTI or kidney infection. inadequate diet	and will it assist to eliminate the problem? Need to get full patient Hx What are the recommended treatments for UTI & kidney infections/stones? What is the possible cause of symptoms (i.e. environmental, dietry, medical)? What fluid intake has the patient had (including type of fluid)? Who is her GP? How often has she felt this way? What has happened in the past? What are the circumstances behind the patient not filling the GP generated prescription? How do I interpret the pathology reports?	Textbook Community Health Information Other Health professionals Journals Complete vital signs assessment Complete MSU, Bloods & Pregnancy test Pain assesment & get pain under control Find out patient Hx Thorough assessment of the patient including vital signs and conversation regarding medical history. Research course textbooks, journals, medical and pathology dictionaries

Fig. 2. Example of FILA and Student Responses.

Contributor's name	Reference	Upload your notes	Feedback		
			Good	Needs improvement	Name and date of person giving this feedback
Pippy Longstocking	ABC (2013)		Used the correct referencing	It is not clear how the values are relevant to this problem	Henry Octopus
			Detailed notes on the values of ANMC	The notes on this learning issue appear to be incomplete	Abby Barramundi

Fig. 3. Research Wiki.

ePBL Step 4: Second Online Synchronous Tutorial

In the second online synchronous tutorial, discussions about new information and ways to manage the problem occur. Students relate their own views and the outcomes of their research. High-quality evidenced-based information from the research wiki (Fig. 3) is referred to and used. At this stage, the tutor facilitates the construction of a concept map as a scaffold to summarise and integrate information gathered from the research stage. Again, while a concept map is relevant to this particular course, as is the wiki, of course other 'products' may be produced where deemed appropriate and befitting the particular discipline(s) context in which the process occurs.

An example of the concept map pertaining to this case study's problem – Jessica Smith, a 25-year old female who presented to A & E with urinary symptoms and a fever – is depicted in Fig. 4. The figure illustrates a nursing assessment branch (one of many branches of a complex concept map) providing information about the patient's nursing care.

ePBL Step 5: Student Reflections and Feedback

The last step is about students evaluating their own performances in problem-solving and reasoning, and the group's overall performance in

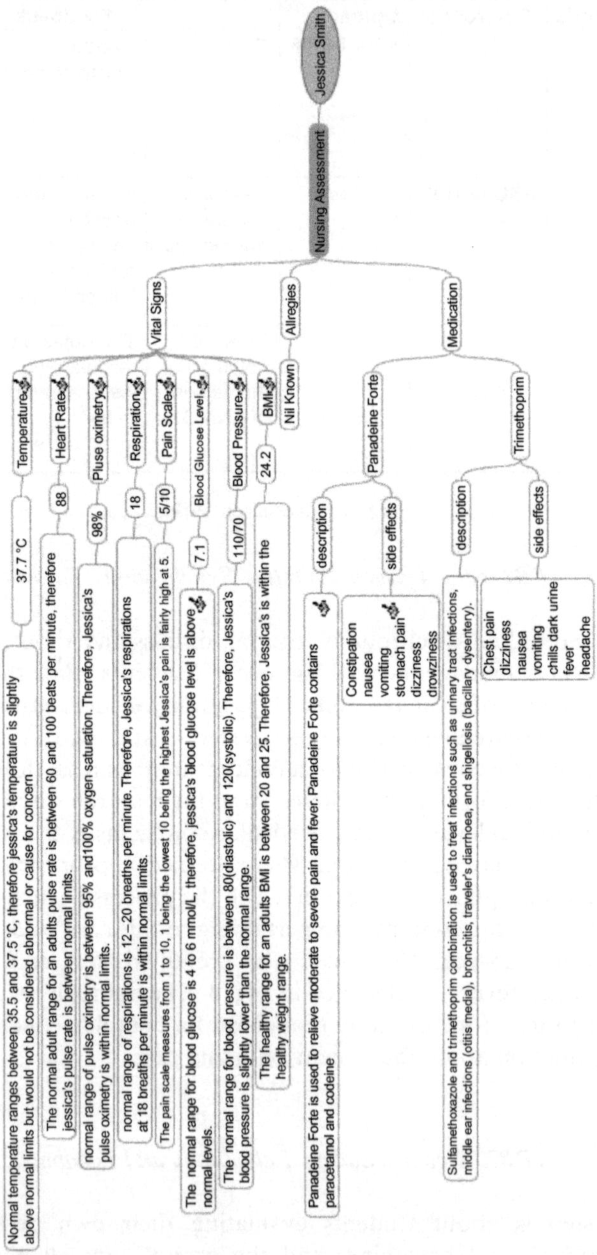

Fig. 4. A Branch of a Concept Map for Learning the Problem.

Contributor's Name	Reflections on your own performance		Reflections on the group's overall performance in managing the problem	Feedback (sign off with your name and date)
	Problem-solving/reasoning	Research		

Fig. 5. Student Reflections and Feedback Wiki.

managing the problem. Students use the final wiki (shown in Fig. 5) to post their reflections and feedback. This also serves as a collegial thank you. The last step (celebration and transition) is often overlooked, but should not be neglected. It is a crucial element of meta-cognition and transformative learning.

CONCLUSION

We have argued in this chapter that PBL has a lot to offer in contemporary blended learning environments, or what we have referred to (following Savin-Baden, 2014) as 'liquid learning' environments. Such environments require agility and flexibility, as well as a whole range of other '21st century skills', such as critical thinking and teamwork. While in its original guise, PBL was very much based in a disciplinary context (medical education), its key strength lies in its ability to develop transferable skills as well as (inter) disciplinary knowledge, both of which are crucial 21st century skills. We have also argued that leveraging the potential of PBL to enhance IBL environments requires on the one hand a strong understanding and ability to apply the original principles underlying PBL, and on the other hand an open mind to adjust PBL practice in such a way that it fits the intended purpose (i.e. the learning outcomes). We have used a limited case study of an online nursing course, to demonstrate the process of applying PBL essentials in practice in a PBL online learning environment of a first-year foundational nursing course that is delivered via on-campus and online modes. Of course this case study applies the PBL essentials on a relatively small scale, so the next step is to implement this across a programme or across a series of courses. A carefully mapped outset of learning outcomes, and an appropriate combination of problems-, projects- and inquiry-based activities, will provide a 'liquid learning' environment that will ultimately prepare confident graduates who will be able to take full advantage of the 21st century learning and professional contexts in which they find themselves.

REFERENCES

Ascroft, K., & Palacio, D. (1996). *Researching into assessment and evaluation in colleges and universities: Practical research series*. London: Kogan Page Limited.

Barrows, H. S., & Tamblyn, R. M. (1980). *Problem-based learning: An approach to medical education*. New York, NY: Springer Publishing Company.

Biggs, J., & Tang, C. (2011). *Teaching for quality learning at university* (4th ed.). Maidenhead: Open University Press.

Biggs, J. B. (1999). What the student does: Teaching for enhanced learning. *Higher Education Research and Development*, *18*(1), 57–75.

Biggs, J. B. (2001). Enhancing learning: A matter of style or approach? In R. J. Sternberg & L. F. Zhang (Eds.), *Perspective on thinking, learning, and cognitive styles* (pp. 73–102). Mahwah, NJ: Lawrence Erlbaum Associates, Inc.

Boud, D., & Feletti, G. (1998). *The challenge of problem-based learning*. London: Kogan Page.

Bradwell, P. (2009). *The edgless university: Why higher education must embrace technology*. London: Demos.

Brown, J. S., Collins, A., & Duguid, P. (1989). Situated cognition and the culture of learning. *Educational Researcher*, *18*(1), 32–42.

Davies, M. (2012). Can universities survive the digital revolution? *Quadrant Online*. Retrieved from http://www.quadrant.org.au/magazine/issue/2012/12/can-universities-survive-the-digital-revolution. Accessed on March 8, 2013.

Deakin-Crick, R. (2007). Learning how to learn: The dynamic assessment of learning power. *The Curriculum Journal*, *18*(2), 135–153.

Deignan, T. (2009). Enquiry-based learning: Perspectives on practice. *Teaching in Higher Education*, *14*(1), 13–28. doi:10.1080/13562510802602467

Halpern, D. F. (1998). Teaching critical thinking for transfer across domains. *American Psychologist*, *53*(4), 449–455.

Hmelo-Silver, C. E., & Barrows, H. S. (2008). Facilitating collaborative knowledge building. *Cognition and Instruction*, *26*, 48–94.

Hung, D. (2002). Situated cognition and problem-based learning: Implications for learning and instruction with technology. *Journal of Interactive Learning Research*, *13*(4), 393–414.

Hung, W., & Loyens, S. M. M. (2012). Global development of problem-based learning: Adoption, adaption, and advancement. *The Interdisciplinary Journal of Problem-Based Learning*, *6*(1), 4–9.

Hutchings, W. (2007). *Enquiry-based learning: Definitions and rationale*, pp. 1–38. Retrieved from http://www.ceebl.manchester.ac.uk/resources/papers/hutchings2007_definingebl.pdf

Jonassen, D. H. (1991). Objectivism versus constructivism: Do we need a new philosophical paradigm? *Educational Technology Research and Development*, *39*(3), 5–14.

Justice, C., Rice, J., Warry, W., Inglis, S., Miller, S., & Sammon, S. (2007). Inquiry in higher education: Reflections and directions on course design and teaching methods. *Innovative Higher Education*, *31*(4), 201–214.

Kahn, P., & O'Rourke, K. (2005). Understanding enquiry-based learning. In T. Barrett, I. Mac Labrhainan, & H. Fallon (Eds.), *Handbook of enquiry and problem-based learning* (pp. 1–12). Galway: CELT.

Kek, M. Y. C. A., & Huijser, H. (2011). The power of problem-based learning in developing critical thinking skills: Preparing students for tomorrow's digital futures in today's classrooms. *Higher Education Research and Development*, *30*(3), 317–329.

Levy, P., & Petrulis, R. (2012). How do first-year university students experience inquiry and research, and what are the implications for the practice of inquiry-based learning? *Studies in Higher Education*, *37*(1), 85–101.

Manista, F. C., & Gillespie, M. P. (2011). Using enquiry-based learning methods to teach Finnegans Wake to undergraduates. *Arts and Humanites in Higher Education*, *10*(1), 85–98.

Mantzoukas, S. (2007). Reflection and problem-enquiry-based learning: Confluences and contradictions. *Reflective Practice*, *8*(2), 241–253.

Marra, R. M., Jonassen, D. H., Palmer, B., & Luft, S. (2014). Why problem-based learning works: Theoretical foundations. *Journal on Excellence in College Teaching*, *25*(3&4), 221–238.

Marton, F., & Saljo, R. (1976). On qualitative differences in learning – I: Outcome and process. *British Educational Research Journal*, *46*(1), 4–11.

Martyn, J., Terwijn, R., Kek, M. Y. C. A., & Huijser, H. (2014). Exploring the relationships between teaching, approaches to learning and critical thinking in a problem-based learning foundation nursing course. *Nurse Education Today*, *34*(5), 829–835. Retrieved from http://dx.doi.org/10.1016/j.nedt.2013.04.023

Pask, G. (1976). Styles and strategies of learning. *British Journal of Educational Psychology*, *46*, 128–148.

Paul, M. (2010). How to organise the transition from a traditional curriculum to a PBL curriculum. In H. van Berkel, A. Scherpbier, H. Hillen & C. van der Vleuten (Eds.), *Lessons from problem-based learning*. Oxford: Oxford University Press.

Pea, R. D. (2004). The social and technological dimensions of scaffolding and related theoretical concepts for learning, education, and human activity. *Journal of Learning Sciences*, *13*(2), 423–451.

Piaget, J. (2006). Reason. *New Ideas in Psychology*, *24*, 1–29.

Prince, M., & Felder, R. (2007). The many faces of inductive teaching and learning. *Journal of College Science Teaching*, *36*(5), 14–20.

Prince, M. J., & Felder, R. M. (2006). Inductive teaching and learning methods: Definitions, comparisons, and research bases. *Journal of Engineering Education*, *95*(2), 123–138.

Rotherham, A. J., & Willingham, D. T. (2010). 21st century skills: Not new but a worthy challenge. *American Educator*, *34*(Spring), 17–20.

Savin-Baden, M. (2014). Using problem-based learning: New constellations for the 21st century. *Journal on Excellence in College Teaching*, *25*(3&4), 197–219.

Scardamalia, M., & Bereiter, C. (2006). Knowledge building: Theory, pedagogy, and technology. In K. R. Sawyer (Ed.), *The cambridge handbook of the learning sciences* (pp. 97–118). New York, NY: Cambridge University Press.

Schon, D. A. (1983). *The reflective practioner: How professionals think in action*. New York, NY: Basic Books.

Sharples, M., Adams, A., Ferguson, R., Gaved, M., Mcandrew, P., Rienties, B., & Whitelock, D. (2014). *Innovating pedagogy 2014: Open university innovation report* (p. 3). Milton Keynes: Open University.

Spronken-Smith, R., & Walker, R. (2010). Can inquiry-based learning strengthen the links between teaching and disciplinary research? *Studies in Higher Education*, *35*(6), 723–740.

Thomas, D., & Brown, J. S. (2011). *A new culture of learning: Cultivating the imagination for a world of constant change*. Lexington, KY: Createspace.

Vygotsky, L. S. (1978). In M. Cole, V. John-Steiner, S. Scribner, & E. Souberman (Eds.), *Mind in society: The development of higher psychological processes*. Cambridge, MA: Harvard University Press.

Wee, L. K. N., & Kek, M. a. Y. C. (2002). *Authentic problem-based learning*. Singapore: Prentice-Hall.

Wood, D., Bruner, J. S., & Ross, G. (1976). The role of tutoring in problem solving. *Journal of Child Psychology and Psychiatry*, *17*(2), 89–100.

ABOUT THE AUTHORS

Amelia J. Ahern-Rindell is an Associate Professor of Biology at The University of Portland. She received her B.S. in Biology from the University of Illinois, Champaign-Urbana; her M.S. and Ph.D. in Genetics and Cell Biology from Washington State University; and was an NIH Postdoctoral Fellow at the Center for Molecular Genetics at the University of California, San Diego. She is involved in biology curriculum modernization and assessment, especially in the area of undergraduate research and scholarship. She has an active research program in which she collaborates with students on the cell and molecular characterization of an animal model of a fatal, genetic disorder. She has mentored ~100 students over the past 20+ years and teaches courses in Cell and Molecular Biology, Genetics, and Evolution. She is a former Department Chair and current President of the Council on Undergraduate Research 2014–2015 after 13+ years as a member of the non-profit organization.

Teresa Bettencourt, Ph.D. in Didactic, is Assistant Professor in the Department of Education of the University of Aveiro, Portugal. She is the General Editor of the Online Review *Indagatio Didactica.* She is the Director of two Master Courses, including the Master in Biology and Geology Teaching – 3rd Cycle of Basic Schooling and Secondary Schooling. She is also the Coordinator of the Supervised Teaching Practice of the University of Aveiro. She belonged to the Advisory Group of Portugal within the European 7th Framework Programme Project INQUIRE. She is an international evaluator of post degrees courses in Latin-America. Her research interests are focused in the educational exploration of 3D virtual worlds, like Second Life®, and in others innovative approaches to science teaching, like Inquire-Based Science Education (IBSE) or flipped-classroom.

Patrick Blessinger is the founder, executive director, and chief research scientist of the International Higher Education Teaching and Learning Association and an Adjunct Associate Professor at St. John's University

(NYC). Patrick is a Fulbright Senior Scholar and Governor's Teaching Fellow and he is a researcher and authority in the areas of leadership, innovation, student engagement, faculty development, learning communities, and global education. Patrick has co-edited and co-authored many textbooks and articles in these areas. Patrick is the editor of two international academic journals, the series editor of a book series in teaching and learning, and the series editor of an anthology series on contemporary teaching and learning. Patrick has taught over 100 college and university courses in management, leadership, education, economics, and technology, and he has managed academic programs at several colleges and universities in the United States and Europe.

Barb Bloemhof has been teaching in higher education using self-directed learning methods for a decade, and is an expert on the use of problem-based learning and inquiry in economics. In addition to the scholarship of higher education, her disciplinary research areas are government institutional arrangements in energy, health, and international economics. She lives and works in Hamilton, Ontario (Canada).

Jacqueline Broerse is Professor of 'Innovation and Communication in the Health and Life Sciences, in Particular Addressing Issues of Diversity and social Inclusion' at the Athena Institute, VU University, Amsterdam. She holds a Master's in Biomedical Sciences (cum laude) and obtained her doctorate in 1998 at the VU University with a thesis on participatory approaches to research priority setting in developing countries. Her research is focused on methodology development for realizing a science-society dialogue in new and emerging (system) innovations in the health and life sciences. She is a member of the Board of the Research School Science, Technology and Modern Culture (WTMC). She coordinates and lectures various courses on health policy and health communication. In addition, she is a member/chair of several educational boards of the School of Life Sciences at the VU University. Furthermore, she is a member of the International Advisory Board of the Dutch Royal Tropical Institute and a member of the working group patient participation of the Dutch Clinical Trial Foundation. Furthermore, she has been appointed Principal Investigator at the Centre for Society and Genomics.

Mindy Capaldi is a Kentucky native who attended Georgetown College in Georgetown, KY, and graduated with degrees in mathematics and history. She left her home state to attend North Carolina State University, where she received a Master's and Ph.D. in Mathematics. Her dissertation was in the field of Algebraic Topology, but she now studies Mathematics Education. After graduate school, Mindy moved to northwest Indiana with her husband, Alex Capaldi. They work together in the Mathematics Department at Valparaiso University. In addition to mathematics education research, Mindy enjoys learning about and implementing innovative teaching strategies, such as inquiry-based learning and flipping classes. When she is not thinking about teaching and math, Mindy likes to read, knit, and play with her son, Henry.

John M. Carfora is Associate Provost for Research Advancement and Compliance at Loyola Marymount University in Los Angeles. John holds graduate degrees from a number of universities, including The London School of Economics, Harvard University, and a Doctorate from Teachers College, Columbia University. A recipient of several international awards, John received the Distinguished Service Award from the National Council of University Research Administrators, and was an IREX Fellow to Russia and a Fulbright Scholar to Ireland. Dr. Carfora is a member of "I-Group" – a National Academy of Sciences and Government-University-Industry Research Roundtable committee on international research – which published the book Examining Core Elements of International Research Collaboration (National Academies Press, 2011). John co-authored *The Art of Funding and Implementing Ideas: A Guide to Proposal Development and Project Management* (Sage, 2011), and wrote the Foreword to *Universitas: The Social Restructuring of Higher Education in America* (Praeger, 1998). He co-authored a popular article on the New Deal economist Stuart Chase (Harvard Magazine, 2004), and wrote Navigating between Teaching, Learning and Inquiry (HETL, 2011). John regularly consults with colleges and universities in the United States and abroad on academic, interdisciplinary, and international research collaborations; comparative education; adult and continuing education; curriculum and instructional development; and leadership and governance in higher education.

Lorraine Carter is the Director of the School of Nursing and of the Centre for Flexible Teaching and Learning at Nipissing University in North Bay, Ontario. She is also a Faculty Member with the Northern Ontario School of Medicine and the School of Rural and Northern Health at Laurentian University. Lorraine has been a champion of adult education and online education throughout her career; her present research interests include the impact of e-learning by health professionals on practice; interprofessional team functioning; leadership; and mobile learning. Lorraine is the past President of the Canadian Network for Innovation in Education (CNIE) and the incoming President of the Canadian Association of the University Continuing Education (CAUCE).

Fabio Dovigo received his Ph.D. in Sciences of Education from Padua University (2000) and his MS in Educational Sciences from Verona University (1996). He is Tenured Associate Professor for Educational Research, Organizational Design and Research Methods at Bergamo University. He is also Director of Post-Graduates Programs, Deputy Director of the School of Management, and Co-ordinator of International Relations at the Department of Human and Social Sciences. Dovigo's current research interests include Research Methods in Education and Organization and Inclusive Education.

Richard R. Hake, after receiving a Ph.D. in Physics from the University of Illinois in 1955, researched high-magnetic-field superconductivity at the North American Aviation Science Center in California and then, starting in 1970, at Indiana University as a Professor of Physics. There, Hake also was assigned teaching responsibilities and became aware of the manifest failure of traditional university pedagogy to promote conceptual understanding. Hake is a Fellow of the American Physical Society and has (a) published over 80 peer-reviewed articles on superconductivity, magnetism, and education; (b) served on the editorial boards of the *Physics Teacher*, *Journal of MultiDisciplinary Evaluation*, and *Journal of Applied Research in Higher Education*; (c) been an advisor for NSF Education Reform Programs at the University of Dallas, Harvard, and Michigan State. Since retirement in 1995, Hake has attempted to promulgate the "interactive engagement" methods that have been relatively successful in undergraduate physics.

Philip Hawes is the Clinical Laboratory Manager at Whitireia New Zealand and is responsible for the implementation of simulation throughout the organization. Philip has a Postgraduate Diploma in Nursing and is a Masters candidate. He is well known amongst the simulation community within New Zealand and overseas. He has spoken at conferences both in New Zealand and abroad.

Henk Huijser is a Senior Lecturer, Flexible Learning and Innovation at Batchelor Institute of Indigenous Tertiary Education. He has a background and Ph.D. in Media and Cultural Studies and has also published widely in the field of learning and teaching in higher education. For more information please visit: http://henkhuijser.webs.com/

Megan Y. C. A. Kek is currently a Senior Lecturer at the Academic Services, University of Southern Queensland (USQ). Prior to joining USQ, Megan's academic work was largely based in Singapore, focusing on marketing education and problem-based learning. She was a team recipient of The Enterprise Challenge Award for excellence in curriculum innovation from the Singapore Prime Minister's Office in 2001 and 2003. The award recognized the most outstanding team whose project has created the highest new value to the public service. Her research interests include educational environments, student learning, problem-based learning, academic development, and use of student data. She writes and presents papers at international and regional conferences. She co-authored a book, *Authentic Problem-Based Learning: Rewriting Business Education* and is currently co-authoring a book with Dr. Henk Huijser. She actively reviews for a range of quality international journals. Megan holds a Ph.D. in Education from the University of Adelaide.

Les Kirkup is a Professor in the Faculty of Science at the University of Technology, Sydney, Australia. He has published widely in peer-reviewed journals throughout his career and has written several textbooks, including *Data Analysis for Physical Scientists* published by Cambridge University Press in 2012. Many of Les' educational development activities have focused on enhancing the student experience in laboratories. His contributions to teaching and learning have been recognized through the award of two national fellowships: in 2007 he was awarded a Carrick Associate Fellowship and in 2011 he was awarded an Australian Learning and Teaching Council National Teaching Fellowship. The latter fellowship was entitled: *Inquiry-oriented learning in science: Transforming practice through forging new partnerships and perspectives.* In 2014, he was awarded the Australian Institute of Physics Education Medal in recognition of his national contributions to teaching and learning in physics.

Wanda S. Konijn is a Lecturer and Researcher at the Athena Institute, VU University, Amsterdam. She graduated in Management, Policy Analysis and Entrepreneurship at the VU University in 2008 and since then she worked at the Athena Institute. Her research interests are developments within education with a special interest in innovations in education. In addition to research, she is involved in teaching and assisting in coordination of several courses concerning, among others, research methodology, policy, and innovation management.

Natalie Kuldell is an Instructor in the Department of Biological Engineering at MIT as well as Founder and President of the BioBuilder Educational Foundation. She graduated from Cornell University with a BA in Chemistry and then received her Ph.D. from Harvard University. After a Post-Doctoral Fellowship at Harvard Medical School, she joined the faculty at Wellesley College where she taught and developed curriculum in the Department of Biological Sciences. In 2003, she was recruited to MIT as they were launching a new undergraduate major and a new department in Biological Engineering. Her leadership in curriculum development and undergraduate education helped position MIT's program as a prime example of interdisciplinary engineering education, particularly in the area of synthetic biology.

Jeroen Maas is fascinated by innovation in relation to personal and organizational development and learning. He approached these topics from different paradigms adapted from both sciences and humanities, and first integrated these different ways of thinking when he started lecturing during his studies. He subsequently worked as a self-employed consultant in Kenya. Interest in the public domain made him return to the Netherlands, where he became Policy Writer at the Dutch Ministry of the Interior and Kingdom Relations. He later returned to the private sector, joining Atos Consulting as a management consultant. He was involved in the introduction of innovative techniques and procedures, and in designing and implementing organizational change and learning processes. In his current activities, he increasingly focuses on the health care sector, teaching two topics close to his heart: management in health care and management of innovations – while searching for innovative methods to get most out of these courses.

Gina J. Mariano is an Assistant Professor of Educational Psychology in the Division of Psychology at Troy University. She teaches educational psychology courses for elementary and secondary teachers as well as psychology courses for undergraduate psychology majors. Her research interests include student learning, metacognition, and knowledge transfer in college students. Her research also examines faculty development among university instructors and its relationship to student learning.

William F. McComas is Parks Family Endowed Professor in Science Education at the University of Arkansas and Director of the Project to Advance Science Education (PASE). He teaches classes in educational research methods, curriculum design, the history and philosophy of science for science educators, issues in science education, and advanced science teaching methods. He earned B.S. degrees in Biology and Secondary Education, M.A. degrees in Biology and Physical Science, and Ph.D. in Science Education from the University of Iowa. He has written and edited several books including *The Nature of Science in Science Education: Rationales and Strategies*. Recently, he has been named the Editor of the American Biology Teacher. Research interests include the improvement of laboratory instruction, effective evolution education, the interaction of the philosophy of science and science teaching, science education for gifted students, and science instruction in museums and at field sites.

Rudolph Mitchell, after holding research and faculty appointments at Boston University, Harvard Medical School, and Brown University, joined the Teaching and Learning Laboratory as the Associate Director for Assessment and Evaluation in January 2004. An educational researcher with expertise in learning behavior, survey research, curriculum development, and program evaluation, Dr. Mitchell is the author of the Cognitive Behavior Survey Series, a collection of surveys that examine learning and related cognitive behaviors of medical students and residents. The surveys have been used by more than 40 medical schools. At MIT Dr. Mitchell has conducted more than 50 studies of curriculum revisions, educational innovations, or new academic programs designed and implemented by faculty members which relate to undergraduate, graduate, or post-graduate training. He has authored or co-authored articles that have appeared in *Academic Medicine*, *Journal of Chemistry Education*, *Journal of Science Education and Technology*, and *Physical Review Special Topics – Physics Education Research*.

Vanessa R. Pitts Bannister is an Assistant Professor and Coordinator of Mathematics Education at the Florida Agricultural and Mechanical University (FAMU). Recently, she was the PI of a Helios Grant program that aimed to improve STEM education through innovative integration and application of proven strategies through coursework and/or professional development. Before this grant, she completed a NSF-funded study of pre-service secondary mathematics teachers' interactions with reform curriculum materials in mathematics methods courses. This line of work resulted in a co-edited book and other peer-reviewed publications. Her research interests include teacher and student knowledge in areas of algebra and rational numbers, teachers' pedagogical and content knowledge with respect to curriculum materials, and equity and diversity issues in mathematics education.

Vincenza Rocco received her Ph.D. in Human Capital Development and Labour Relations from Bergamo University (2014) and her MS in Pedagogical Studies from Bergamo University (2010). She is Research Fellow in Educational Research at the University of Bergamo. She was involved in the European project INQUIRE (Inquiry-based teacher training for a sustainable future) in order to contribute to evaluation of Inquiry Based Science Education method. Rocco's current research interests include Environmental Education, Evaluation, Research Methods in Education, and Inclusive Education.

Richard E. Shope is a Science Educator and Community Science Research Fellow at the Center for Urban Resilience at Loyola Marymount University, and he is Director of EcoVoices, a research, education, and workforce development program for urban youth at the San Gabriel River Discovery Center. With over 40 years of experience as a poet, mime artist, and educator, Shope is known for his abilities to communicate science through the performing arts. He coordinated nationwide education and public outreach programs for various flight projects at NASA's Jet Propulsion Laboratory, focused on reaching urban youth and rural American Indian youth. His Mimewrighting program has reached over one million schoolchildren throughout Southern California. He earned a B.A. degree at Antioch University Los Angeles; M.S.Ed. and Ed.D. degrees in Science Education at the University of Southern California, chaired by William F. McComas. Research/Practitioner interests include inquiry-driven informal science education and teaching for conceptual change.

Susana Silva has a degree in Geography, specialization in Environment and Development, Faculty of Letters, University of Coimbra; has a Masters in Human Geography, Spatial Planning and Development, Faculty of Letters, University of Coimbra; is a Doctoral Student in Human Geography, Faculty of Letters, University of Coimbra; is a Researcher at the Centre of Geography and Spatial Planning Studies (CEGOT) from the Universities of Coimbra, Porto, and Braga.

Lorraine Stefani is a Professor of Higher Education Strategic Engagement at the University of Auckland, New Zealand. Lorraine has been, and continues to be involved in many strategic educational initiatives including: President of HETL; Member of the Education Advisory Group for the New Zealand Royal College of General Practitioners (RNZCGP); Director of the Northern Regional Training Hub (NoRTH) for Health Workforce New Zealand; Member of Inaugural Board of Directors for Ako Aotearoa

(NZ National Centre for Tertiary Teaching Excellence); Member of Advisory Board of Ako Aotearoa; Member of Inaugural Board of Directors for the Institute of Learning and Teaching in HE (UK); Chair of the Personal Development Planning in HE (Scotland Group). She is a member of many professional societies and associations including Senior Fellow of the Staff and Educational Development Association (SEDA UK), Senior Fellow of the Higher Education Academy (HEA UK), and member of the International Leadership Association (ILA).

Ana Cristina Tavares, Ph.D. in Biology, works at University of Coimbra (UC), is a member of the Research Centers of Pharmaceutical Studies (CEF-UC) and Human Development (CEDH-UCP). At UC, she was teaching assistant in Botany Department (1981–1992) and Faculty of Pharmacy (2002–2003); she worked at Coimbra Botanic Garden (COIBG) (1997–2014) and was responsible for Education Service (1997–2012). She holds International Diploma "Education in Botanic Gardens" (Kew Gardens, London, 2002) and Graduation "Plants Medicines and Healthcare" (Faculty of Pharmacy, UC, 2007). She is responsible for COIBG as Partner in INQUIRE European Educative Project (2011–2013) – 7th FP – she belonged to the Portuguese Inquire Advisory Group. She published refereed papers and books: *Discovering the world of plants* (2011), Lisbon; at UC Coimbra Press: *Medicinal and Aromatic Plants at COIBG* (2010), *Rhythms of COIBG* (2012) and *The alga who wanted to be flower* (2013), this in five bilingual versions; recently, on-line Eumed.net, *Education in Botanic Gardens-16 years of experience* (2015).

Ruth Terwijn is a Lecturer of Nursing at USQ, Australia. Her interest in problem-based learning was first established when teaching first-year student nurses. There was a need to develop authenticity in learning to augment undergraduate critical thinking to the application of practical skills. In consultation with Dr. Megan Y. C. A. Kek, Barrows' model was used as a transformative learning and teaching intervention to change how students learn and how lecturers can teach in the course. This model, started as a face-to-face model, was successfully adapted for an online blended learning environment, culminating to an ePBL model, keeping Barrows' PBL principles and essentials as a curriculum and pedagogy. It has provided a sustainable and practical platform for introducing nurses to the clinical milieu. Ruth is currently completing her Ph.D. looking at the experience of undergraduate nursing students choosing to study outside their homeland at an English speaking university.

Wendy Trimmer is Academic Leader for Postgraduate Nursing and Health Programmes in Te Kura Hauora, the Faculty of Health at Whitireia New Zealand. Wendy is a Registered Nurse and has a Masters (Applied) in Nursing. Wendy is the champion in implementing Context Based Learning throughout the Faculty of Health, and has presented in both national and international conferences on the topic.

Dylan Powell Williams' research is focused on the development and evaluation of innovative approaches to chemistry teaching primarily based on the Context/Problem Based Learning (C/PBL) approach. In June 2012, Williams was awarded a University Teaching Fellowship in recognition of his innovative approach to teaching. In 2012, Williams led the production of four new peer-reviewed C/PBL resources. Williams currently (2014–2015) lead a Royal Society of Chemistry funded project to develop a new C/PBL module which will engage undergraduate chemists in industrial research problems. Williams is a Member of both the Royal Society of Chemistry (RSC), a Senior Fellow of the Higher Education Academy and Committee Secretary of the RSC Tertiary Education Group.

Baiba Zarins is a Member of the University Health Network (UHN) collaborative academic practice team. Baiba's portfolio supports UHN's commitment to innovation in health care education and professional development. Baiba facilitates an innovative second entry Bachelor of Science in Nursing (BScN) program in collaboration with Nipissing University and four Toronto-based academic health care delivery organizations, utilizing the tenants of narrative inquiry and cognitive apprentice pedagogies. Other roles include facilitation of UHN's organizational capacity and knowledge networking with local and international professional development opportunities; all of which foster unique responsiveness to health human resource needs. Past collaborations with the Registered Nurses Association of Ontario, within the Best Practice Guidelines (BPG's) and Advanced Clinical Practice Fellowship programs, sustain evidence-informed practice initiatives. Baiba holds a Masters of Health Studies from Athabasca University and is currently enrolled in the doctoral program at the Fielding Graduate University in Human and Organizational Systems. Her 25-year nursing career spans many roles within clinical, educational, and administrative capacities in both acute and community care sectors. Baiba holds leadership positions on the Board of Directors of Kristus Darzs Latvian Home, a long-term care facility located in Woodbridge, Ontario.

Marjolein B. M. Zweekhorst is Associate Professor at the Athena Institute, VU University, Amsterdam. She graduated in Medical Biology at the VU University, Amsterdam (1996). After her graduation, she joined the Department of Biology and Society as a researcher. She then became a Ph. D. student and wrote her Ph.D. thesis on institutionalization of an interactive approach to technological innovation (defense 2004). She became Mastercoordinator of Master Management Policy-Analysis and Entrepreneurship in the Health and Life Sciences in 2005, and since 2013, she is the Program Director of this program. Next, she is Vice-Chair of the Examboard of Biomedical Sciences, Chair of the Educational Board of Health Sciences, and Member of the Curriculum Committee of the Bachelor Program Health and Life Sciences of the School of Life Sciences, all at the VU University. Furthermore, she is member of the International Academic Board of the Dutch Royal Tropical Institute. Marjolein has a particular interest in quality improvement in higher education. Her current research is focused on methodology development for interactive policy and technology development in the field of (international) public health, biotechnology and biomedical sciences, and on the most effective and efficient way to train master students in this field of research.

AUTHOR INDEX

SUBJECT INDEX